Genetics for
Medical Students

Genetics for
Medical Students

E. B. FORD, F.R.S.

Professor in the University of Oxford

CHAPMAN AND HALL

LONDON

First published 1942 by Methuen & Co. Ltd.
Second edition, 1946
Third edition, 1948
Fourth edition, 1956
Fifth edition, 1961
Sixth edition, 1967
Seventh edition, 1973, published by Chapman and Hall Ltd.,
11 New Fetter Lane, London, EC4P 4EE

© 1973 E. B. Ford

ISBN 0 412 10950 6

Printed in Great Britain
by Richard Clay (The Chaucer Press) Ltd.,
Bungay, Suffolk

Distributed in the U.S.A.
by Halsted Press, a Division
of John Wiley & Sons, Inc., New York

Library of Congress Catalog Card Number 73-13385

Contents

CONTENTS

Preface

The importance of genetics is being widely recognized by the medical profession. Practitioners require a knowledge of this subject in the course of their work, and students are faced with it in examinations. Yet though textbooks of genetics are available in great variety, the number of them which treat of human heredity is small. Indeed a work providing a brief and elementary introduction to that subject has long been required, and this book is an attempt to meet that need.

The scope of this account has to some extent been determined by the experience gained during many years of teaching genetics to medical students at Oxford. I have made no attempt to present a compendium of knowledge on the subject. On the contrary, the book is designed to illustrate its chief principles. Consequently, the various inherited conditions which are described have been chosen as illustrative instances only. However, a classified list of some of the more important, or well authenticated, of the genes known in man is provided on pp. 210–18. Further information on them should be obtained from works devoted to special branches of human genetics. Among these, one is of such outstanding merit that I wish to draw attention to it here: I refer to E. A. Cockayne's *Inherited Abnormalities of the Skin and its Appendages* (Oxford Press, 1933).

The examples which I have employed are mainly human, but I have made sparing use of other material where this seemed particularly appropriate and helpful. Thus to describe autosomal linkage without reference to lower forms would greatly embarrass the explanation of a phenomenon which has an important bearing on medical work.

I hope that those who read this book will not be content to end their genetic studies with it. In a brief Bibliography, I have suggested a few works which supply more detailed information on the subjects covered by each chapter, so that students may be able to

PREFACE

extend their knowledge in the direction they desire. To those who
require a fuller general account of human heredity, I would particularly recommend the admirable treatise by J. A. Fraser Roberts,
An Introduction to Medical Genetics (Oxford Press, 1940). The
provision of a Bibliography of this kind greatly reduces the number
of references which it is necessary to supply. Indeed, it is hardly
possible, and certainly not desirable, to quote authorities for all
the statements made in a textbook. In general, therefore, I have
given references only for those which it is likely that students may
wish to expand, particularly if the necessary information cannot be
obtained except from original sources.

This is a work addressed primarily to medical men, consequently
I have assumed that the reader possesses such a knowledge of
cytology as is acquired by medical students at an early stage in
their career. However, I have provided an Appendix giving a brief
account of the relevant aspects of that subject, so that the book can
be read also by those who have no previous acquaintance with it.
Furthermore, much new light has been thrown on the details of
mitosis and meiosis in recent years, and those who suspect that
their information on these processes may be somewhat out of date
are recommended to revise them with the help of the descriptions
given on pp. 201–9.

The need for further research on human genetics will be apparent to those who read this book, and some of the methods appropriate to such investigations are here described. If practitioners will
record any relevant data which they may obtain, they will be able
so to advance this subject that in a short space of time its value will
be greatly increased.

Part of this book was written while I was occupying the position
of Visiting Professor at the Galton Laboratory, at that time moved
from London to Rothamsted Experimental Station, Harpenden,
and Professor R. A. Fisher, F.R.S., has given me the benefit of his
opinion on it throughout. His inspiring analysis of genetics has
placed all students of that subject in his debt, but I owe to him
even more than this. Our close association, maintained for many
years, has been a source of constant encouragement and help in all
branches of my work.

PREFACE

It is very important that a discussion of human genetics should conform to the requirements of the medical profession; nor is its precise scope, judged from that point of view, easy to define. I have been particularly fortunate, therefore, in obtaining the advice of Professor W. E. Le Gros Clark, F.R.S., upon this matter. Indeed, it was at his suggestion that this book was written. It is a pleasure to record my gratitude to him for the very considerable trouble which he has taken in reading and criticizing it. I should also like to express my grateful thanks to Professor E. S. Goodrich, F.R.S., for his help. Professor G. D. Hale Carpenter has been so kind as to read the proofs, and his comments have been of great assistance to me.

Dr G. L. Taylor, of the Galton Laboratory Serum Unit, has given me the benefit of his valuable advice on those sections of this book which deal with blood-grouping. I am most grateful for his help. I am indebted to the Editor of the *Annals of Eugenics* for permission to publish Fig. 9. I should like to express my thanks to Mrs M. Nicholson for the care which she has taken in preparing Figs. 1–8.

It had been my intention to submit each chapter of this book, as it was written, to my cousin, William Rowland Thurnam, M.D., the well-known authority on tuberculosis. The keen interest which he took in its plan was an incentive to write it. His death during the course of the work has been a great loss to me. The advice of one who was eminent not only in the medical profession but as a literary critic and artist of exceptional attainments, would have been of much value in preparing this survey of human genetics.

OXFORD,   E.B.F.
February 1942

Preface to the Fifth Edition

This book remained substantially unchanged until the fourth edition, published in 1956, when Messrs Methuen made it possible for me to bring it up to date and to enlarge it. By far the greatest recent contributions to human genetics have been made in the field of serology and, accordingly, the sections dealing with the blood groups were extended from one chapter to four. I am deeply indebted to Dr R. R. Race, F.R.S., and Dr Ruth Sanger for the helpful guidance and criticism which they gave me on that occasion.

Now that a fifth edition is required, it is still in the serological sections that the principal changes have had to be made, and these have accordingly been adjusted very considerably. Other parts of the book have also been brought into line with recent discoveries.

The notation of the blood groups, as generally used, is still chaotic. Not only has it been out of accord with that universally accepted in genetics, but it has been inconsistent from one blood group system to another, while constant confusion has arisen between genes and antigens. This has greatly increased the difficulty of a subject in itself intricate. I had in 1955 proposed a uniform notation which largely overcomes these defects. This is of course employed here. The failure of serologists to adopt a notation (whether my own or some other) in agreement with the standard usage in all other branches of genetics and consistent in itself is severely retarding the incorporation of their subject into the general field of biology. The longer there is delay in this matter the more difficult will be the changes when these have finally and inevitably to be made.

The blood groups were treated as polymorphisms for the first time in *Genetics for Medical Students*. When the book appeared in 1942, such an approach by no means escaped criticism. However, it received emphatic approval from Sir Ronald Fisher, F.R.S., whose contributions to the theory both of serology and polymorphism have been fundamental. Today, the control of the blood groups as balanced polymorphisms is no longer seriously challenged. It was demonstrated originally by the theoretical arguments outlined in Chapter 5. Though these have, of course, been somewhat developed since the first edition, they have not been radically changed. A logical extension of them led to the suggestion that blood grouping should be undertaken on patients suffering from different types of disease, and that significant differences might then be found between them and the normal populations from which they were derived (Ford, 1945, p. 85). Investigation along these lines has now yielded striking positive results. It may be hoped that these will lead to advances in diagnosis and, conceivably, in treatment. Moreover, they provide observational data establishing the selective importance of the human blood groups, and the validity of the conclusion which led to their treatment as polymorphisms.

Dr R. R. Race, F.R.S., has once again given me most valuable help in certain technical aspects of serology. I am deeply grateful for the very considerable trouble he has taken, but wish to stress that he is in no way responsible for any errors which this book may contain. I am greatly indebted also to Dr P. L. Mollison for information on incompatible transfusions.

1960   E.B.F.

Preface to the Seventh Edition

There are two distinct ways of writing a textbook, and a compromise between them is not usually to be recommended. Either one may attempt to 'cover the subject', as the saying is, or to develop *principles* with examples sufficient to illustrate them but no more. This work is of the second kind though, in view of the need for the more comprehensive introduction to serology required by medical men, it departs somewhat from it, towards a middle course, in describing the blood groups. Yet one has only to compare the account given here with the great work of Race and Sanger (1968) to see how far is this small book from covering that subject. Now, in its seventh edition, it is nevertheless more ambitious than previously. For at its first appearance in 1942 genetics was something of a luxury to the medical student; today it is a necessity.

This is due to the fact that its applications are becoming more apparent, as well as to the growth of the subject. We owe that progress to the labours of medical men working in Hospitals, as Consultants and as General Practitioners, who have collected much information which they are so well fitted to obtain; and, in particular, to research scientists in many countries.

I would like here especially to draw attention to the outstanding advances achieved by the Nuffield Institute of Medical Genetics at Liverpool: an organization fortunate in its endowment from the Nuffield Foundation, in its Staff of distinguished experts and in its Director Professor C. A. Clarke, F.R.S. who combined with that office the post of Professor of Medicine in Liverpool University. With a view so wide as to illuminate human polymorphism from that of butterflies, and the distinction required to become President of the Royal College of Physicians, it was he who established medical genetics at Liverpool upon lines that have proved in the highest degree successful. His own technique of obtaining protection against haemolytic disease of the new-born due to the Rhesus blood group, has received world-wide recognition. In addition, P. M. Sheppard, F.R.S. Professor of Genetics in the University, has constantly contributed knowledge and originality, especially in the studies on the Lepidoptera, to furthering the research programme of the Nuffield Institute.

In spite of his exacting commitments at the Royal College of Physicians during his term of Presidency, Professor C. A. Clarke, in addition to much other help, has drafted four paragraphs for this book (pp. 154–5) giving the most up to date account of his own work on the prevention of Rh immunization of a mother by her foetus. I am deeply indebted to him for his kindness in providing a statement of the highest authority on this important aspect of medical genetics.

I am greatly indebted to a number of other authorities who have been so good as to give me their valuable help in preparing the present edition of this book: Dr M. G. Bulmer, Dr E. R. Creed, Professor C. D. Darlington, F.R.S., Professor D. A. P. Evans, Dr P. T. Handford, Professor K. G. McWhirter, Professor P. M. Sheppard, F.R.S. and Professor D. J. Weatherall. They have been so kind as to take much trouble to place their special knowledge at my disposal.

I have also benefited from discussions with Dr D. R. Lees. He has furthermore been so kind as to undertake the considerable labour of reading the proofs. For, following the wise precept of King George III on this matter, I do not as author trust myself to carry out my own proof-correcting unaided.

December 1972   E.B.F.

CHAPTER I

Mendelian Heredity[1]

I.I MENDEL'S TWO LAWS

The human body, with all its attributes, is a product of heredity and environment. Neither of these two agencies is constant, and changes in them may lead to variations in one or more physical or mental characters. In studying the interaction of two variables it is always desirable to consider each separately before examining their joint effects. Therefore we will first discuss the operation of inheritance, and subsequently analyse the part played by the environment and its interaction with heredity (see Chapter 4).

The characters of the body, in the widest sense, are controlled by hereditary factors (or *genes*) present in every cell. They are passed from parent to offspring in the gametes. These genes therefore exist in pairs, the 'allelomorphs' (or *alleles*), whose members are derived the one from the father and the other from the mother. The two genes composing such allelomorphic pairs may either be similar ('homozygous') or dissimilar ('heterozygous'), and the individuals possessing such allelomorphs are called 'homozygotes' and 'heterozygotes' respectively. The simplest situation therefore is that in which a given character is determined by a *pair* of hereditary factors. Instead of contaminating each other when brought together into the same cell ('blending inheritance', pp. 26–7), the genes retain their own identity ('particulate inheritance'); even the unlike members of a pair of heterozygous allelomorphs remain quite distinct. Fertilization is an additive process; for the gametes are effectively only half a cell each, and they combine to produce the 'zygote', which is the first cell of the new organism. Some mechanism must therefore exist by which the

[1] Those who have no knowledge of Cytology are recommended to read Appendix I (pp. 201–9) before beginning this book.

I

members of each allelomorphic pair of genes are separated from one another and pass into different gametes. These then contain but one member of each of the allelomorphs, the pairs being restored at fertilization.

These conclusions on the nature of heredity were first enunciated by Father Gregor Mendel (1822–84), working at the Augustinian monastery of St Thomas at Brno (formerly known as Brünn) in Moravia, of which he lived to become Prälat. He summarized his results in the form of two laws. The first of these may now be considered.

Mendel's First Law (The Law of Segregation) states that characters are controlled by pairs of genes the members of which separate, or *segregate*, from one another during the formation of the germ-cells and pass into different gametes. The pairs are restored at fertilization, which allows of their recombination in definite proportions. Consequently the characters to which they give rise may also segregate: for they will appear in subsequent generations with definite numerical frequencies.

It will be apparent that this law is concerned with two sets of events. The behaviour of the genes, and that of the characters to which they give rise. Segregation relates fundamentally to the separation of the pairs of genes from one another and their passage into different gametes, but it was inferred from the visible separation, or segregation, of the characters which those genes control. A mating between two organisms which differ from one another in respect of a pair of contrasted characters (p. 38) produces a heterozygous form whose appearance, for our immediate purpose, is immaterial. Mendel found that if two such heterozygotes are crossed, the original types will emerge from the hybrid mixture as pure as they were before: that is, they segregate in the succeeding generation, and in definite proportions.

When Mendel published his results in 1866, no mechanism was known which could effect the segregation of the genes. However, his work was long ignored and only received due attention in 1900, sixteen years after his death. By that time the chromosomes had been discovered, and it was soon obvious that these constitute vehicles perfectly adapted to carry the genes and to ensure their segregation. The parallel is indeed striking between the behaviour of the chromosomes, detected by the microscopic study of cells, and that of the genes, determined by tracing the inheritance of the characters which they control. The genes are present in pairs (allelomorphs), as are the chromosomes (homologous chromosomes). The members of these pairs, both of genes and of chromosomes, are derived respectively from the two parents. Consequent upon Mendelian segregation, the genes constituting the allelomorphs separate from one another and pass into different gametes, as do the members of the homologous pairs of chromosomes, owing to meiosis. The gametes then contain one member only of the pairs both of genes and of chromosomes; but these are restored by the additive nature of fertilization. It will later become apparent that other equally exact parallels exist between chromosomes and genes (for example, those provided by 'crossing-over', pp. 15–24). But the chromosome theory of heredity is not dependent upon mere similarity of behaviour, however striking: that the chromosomes carry the genes is a fact which has now been demonstrated by many critical experiments.[1]

It will at this stage be helpful to take an example of the working of Mendel's first law. Several other consequences of its operation can then easily be appreciated. For this purpose we will trace the inheritance of a peculiar condition of no direct, though of considerable indirect, importance (pp. 121–4). In mankind, about one individual in four is unable to taste the organic compound phenyl-thio-urea and some of its chemical allies in concentrations so low as fifty parts per million.[2] This substance is intensely bitter to those who can detect it, and has the formula:

$$\text{phenyl-thio-urea}$$

The ability to taste it is due to the action of a pair of genes, which may be denoted *TT*. However, they may exist in another

[1] For a survey of this subject, see Sinnott, Dunn and Dobzhansky (1958).
[2] 'Non-tasters' can often detect it in concentrations of 400 parts per million (in aqueous solution).

form (*tt*) responsible for the inability to do so. These two types are both homozygous, for the allelomorphs with which we are concerned are made up of similar members: both *T* or both *t*. They are carried respectively on a pair of homologous chromosomes which separate from each other during meiosis, one member passing into each gamete. This brings about segregation, so that every gamete contains one member only of the genes determining the power to taste phenyl-thio-urea. Thus a 'taster' produces gametes each containing a single *T*, while those of a 'non-taster' each contain *t* only. Should a taster and a non-taster marry, their children will possess pairs of homologous chromosomes once more, whose components are derived from the father and from the mother. These will each carry a gene controlling this particular tasting faculty. Thus the pairs of allelomorphs are restored, but these are heterozygous pairs (*Tt*), composed of *T* from one parent and *t* from the other (Fig. 1).

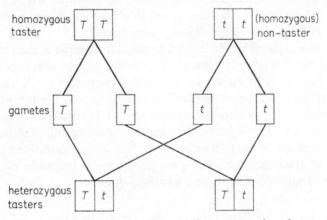

FIG. 1. The production of heterozygotes from a marriage between the two contrasted homozygous types. The characters are the ability to taste phenyl-thio-urea, which is dominant to the inability to do so.

In some conditions the operation of two unlike allelomorphs gives rise to an intermediate effect (p. 215). More often, however, the heterozygotes closely or completely resemble one of the homozygous types. In these circumstances, when the same result is produced in the heterozygote as in one of the homozygotes, the character is called a 'dominant'. The alternative character, which is obscured in the heterozygote and appears only in the other homozygote, is known as a 'recessive'. The inability to taste phenyl-thio-urea is recessive, so that the heterozygotes manifest the dominant condition and are 'tasters'.

Heterozygous individuals, produced by a mating of two homozygotes, are said to constitute the 'first filial' (or F1) generation of the family which is being investigated. The original parents then belong to the 'first parental' (P1) generation. When it is necessary to refer to the grandparents of the F1 generation, these are denoted by P2, and so on backwards. A mating between two individuals of the F1 generation gives rise to a 'second filial' (F2) generation. This cannot normally be studied directly in man, since it arises from a brother and sister marriage. However, its equivalent, for our present purpose, is constantly encountered: a marriage between two heterozygotes. The results to which this gives rise must be considered in some detail.

A heterozygous taster carries the dissimilar allelomorphs Tt. Consequent upon meiosis (see Appendix I), the gametes receive one member only of this pair of genes, T or t. Sperms or eggs carrying one type or the other are thus produced in equality. On a marriage between two such heterozygotes, the chances are equal that a sperm carrying T meets an egg carrying T or t, so producing zygotes of the constitution TT and Tt in equal numbers. Similarly, with the equally numerous sperms carrying t which, meeting the two types of eggs as before, produce Tt and tt zygotes, also in equal numbers. Three types of offspring therefore arise, possessing TT, Tt and tt, in a ratio of $1:2:1$. These are the proportions always produced from the random combination of two types having equally numerous alternative phases: they are not the prerogative of genetics. If we toss together two coins a considerable number of times, our throws give 25 per cent both obverse; 50 per cent, one obverse and one reverse; and 25 per cent both reverse; three groups in a ratio of $1:2:1$, as in the genetic instance. However, in the example which we are studying, the three classes of zygote constituting the F2 generation will not give rise to three distinct classes of individuals: tasters, imperfect-tasters and non-tasters, for, as already mentioned, the first two of these are combined, due to dominance. Consequently two classes only appear, tasters and non-tasters, in a 3:1 ratio, representing a 1:2:1 ratio with the first two terms added together (Fig. 2).

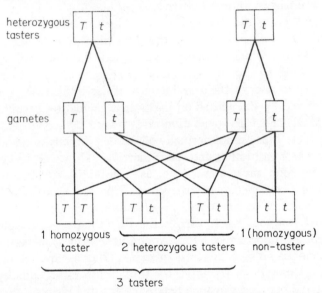

FIG. 2. Segregation among the children of a marriage between two heterozygotes. The characters are the same as in Fig. 1.

In discussing the P1 generation of this example no indication was given whether the taster was the male or the female parent, and none is needed. Save for those instances especially related to the sex-determination mechanism, to be discussed in Chapter 2, the two sexes are exactly equal in the contribution which they make to the heredity of their offspring, and the result is the same however the genetic types or the characters are distributed in regard to sex. It will now be evident, moreover, that dominants can be of two genetic types, homozygotes or heterozygotes, while recessives can be of one type only, for they must always be homozygous.

We have just examined the result of a marriage between two

MENDELIAN HEREDITY

heterozygotes. It is now necessary to consider the highly important situation in which one parent is a heterozygote and the other a homozygote. For this purpose we will use the same character as before, the ability or inability to taste phenyl-thio-urea.

As already indicated, a heterozygous taster bears the genes Tt on a homologous pair of chromosomes, and these separate during meiosis, carrying T into one half of the gametes and t into the other. However, the non-tasters being recessive can only be homozygous, and possess tt; so that all the gametes which they form are similar, and carry t. In a marriage involving these two types it is immaterial which of them is the male and which the female, and the chances are equal that the gametes of the non-tasters, necessarily provided with t, meet gametes bearing T or t produced by the heterozygote. Thus when the pairs of homologous chromosomes, and consequently the allelomorphs, are restored at fertilization, the combinations Tt (tasters) and tt (non-tasters) are produced in equality: heterozygotes and homozygotes appear in a ratio of 1:1. This simple type of mating is called a 'back-cross', because it results when a heterozygote, of the F1 generation, is crossed back to one of the parental forms: but it must not be so defined, for it can arise in other ways. Rather, the following definition should be employed: 'A *back-cross* is a mating between a heterozygote and a homozygote, and it leads to segregation in equality.' This expresses all the essential features of the situation. The individuals produced by a back-cross constitute the R2 generation (Fig. 3).

With an intermediate heterozygote, a back-cross to either homozygote will lead to the segregation of distinct characters. When dominance is complete, however, this result is attained only by the back-cross to the recessive. That to the homozygous dominant produces an apparently uniform R2, resembling both parents, but consisting of homozygotes and heterozygotes in equality. The reality of this segregation, concealed as it is, could be established if the members of such a R2 generation were to mate with recessive individuals. Half of them would then produce the dominant type only, and half would produce dominants and recessives in equality.

7

The class of test just suggested is an important one. For if we wish to determine the genetic constitution of an individual during experimental work on animals and plants, it is always desirable to cross it with the recessive type. The advantage of this is evident, since recessives are of known constitution, being necessarily homozygous, and allow the demonstrable segregation of recessives as well as dominants among their offspring when mated

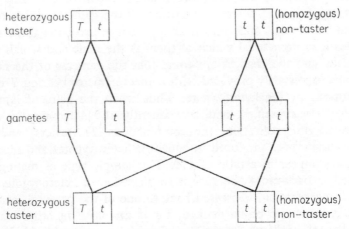

FIG. 3. A back-cross (heterozygote × homozygote), showing that segregation takes place in a ratio of 1:1 among the offspring of this type of marriage. The characters are the same as in Figs. 1 and 2.

with heterozygotes. A test of this kind can, of course, be employed to prove the truth of an assertion made at an earlier stage: that the 3:1 ratio obtained in F2 segregation is, in reality, a ratio of 1:2:1 obscured by dominance. If such segregating dominants be mated with recessives, one third of them will produce dominants only. The remaining two thirds will produce dominants and recessives in equality; this is a back-cross result, and it arises from matings between recessives and the concealed heterozygotes. Clearly the dominant class of the F2 generation is a composite one, including homozygotes and heterozygotes in a ratio of 1:2, as postulated on theoretical grounds.

We have now considered the two types of segregation which

can take place with single-factor inheritance: that characteristic of the F2 generation and of the back-cross. Furthermore, we have seen that it is possible to demonstrate the occurrence in both of them of the fundamental ratios expected, even when these are obscured by dominance. It has already been stressed that one of the most important properties of such segregation is the re-appearance of the original homozygous types among the offspring of heterozygotes. When a marriage occurs between two similar homozygotes, both non-tasters or both homozygous tasters in our example, all the progeny are homozygotes also. Further, the type remains pure indefinitely, until a mating with the heterozygote or with the other homozygote takes place. That is to say, a homo-zygous line always breeds true, whether the genes have passed through a heterozygote or not, and unlike genes do not contaminate one another even when brought together into the same cell. Thus Mendelian heredity has provided an entirely new concept of the hybrid, one of mosaic not of blended type, and of the purity of the hereditary units.

Both the ability and the inability to taste phenyl-thio-urea are very common conditions in the population. However, the study of hereditary disease constitutes an important aspect of human genetics, and here one of the alternative states will be rare. This leads to certain situations which require brief mention.

Friedreich's Ataxia is a rare disorder in which a degeneration of the nervous system produces at first a loss of power in the lower extremities. The patient sways as he stands, and walks with difficulty. Later the ataxia involves the trunk, arms and head, speech becomes difficult, and advanced cases may be unable to sit up. The condition is inherited as a simple recessive and, consequently, it can hardly ever arise except from a mating between two hetero-zygotes. Since the disease usually appears during the second decade of life, those affected by it will certainly reproduce less frequently than normal individuals. Even so, all their offspring will almost invariably be of the normal (dominant) type; for only when they marry a close relative (pp. 188–9) is there any reasonable possibility of a back-cross result, which would show that the ostensibly healthy partner in the union is in reality a heterozygote. Still less is it likely that a marriage between two affected persons will ever be observed. Thus it is clear that those suffering from a rare disease inherited as a simple recessive will almost always have normal parents and, at first sight, the condition will appear merely sporadically in the history of affected families.

Special qualifications are also necessary in considering rare conditions inherited as simple 'dominants'. All affected individuals will then have an affected parent: a direct history of the condition being traceable in one ancestral line in unbroken succession (except for the rare occurrence of mutation, see Chapter 3). Consequently the gene concerned is perpetuated through a series of back-crosses, producing normal and affected heterozygous individuals in approximate equality in each generation. Matings between two heterozygotes, each therefore manifesting the rare disease in question, are liable to be of immense rarity. They are most likely to arise from the marriage of close relations and in those conditions in which the disability does not appear until after the normal age of parenthood. Huntington's Chorea is a simple 'dominant' of this kind, though so far studied only in the heterozygous phase, for the symptoms seldom become noticeable before the age of thirty, and often later. It is characterized by involuntary muscular movements, and mental deterioration progressing to insanity.

It will be clear that the circumstances attending the transmission of a rare dominant disease hardly ever permit its manifestation in the homozygous form. Now the definition of a dominant condition is one in which the effect of the heterozygote is indistinguishable from that of one of the homozygotes. Yet in the vast majority of the rare characters which have been described as simple dominants in man, no opportunity has arisen for comparing them in these phases. Here dominance in reality means no more than that the disease, or other characteristic in question, is unifactorial and is expressed in the heterozygotes. In other animals and plants, in which experimental studies are usually possible, such a situation should not be described as a dominant until the heterozygotes have been compared with both homozygotes and found to agree with one of them. For the occurrence of true dominance is of much evolutionary significance (Chapter 4) and should not be confused with that in which the heterozygous effect is separately detectable. In man, however, it may be extremely difficult or impossible to obtain the information necessary to draw this distinction, while the term 'dominance' has been so widely applied that it might hardly be practicable to attempt severely to limit its use. However, it should always be clearly indicated whether or not the nature of 'dominant' characters has been established. Indeed, it is highly probable that many of those that are so called in man are not, strictly speaking, dominants at all (pp. 84–5).

The analysis so far undertaken has related to the segregation of a single pair of allelomorphs upon the basis of the first law of Mendel. His second law is concerned with the behaviour of two or more pairs of genes when studied together.

Mendel's Second Law (The Law of Independent Assortment) states: When two or more pairs of genes segregate simultaneously, the distribution of any one of them is independent of the distribution of the others.

That is to say, if we consider two characters segregating in a 3 : 1 ratio, this law tells us that nothing interferes with the random distribution of each: there is no tendency for the parental combinations to be preserved in the offspring, nor for the two dominant and the two recessive types to segregate together. Rather, the result will be the mathematical combination of two 3 : 1 ratios; that is to say, a ratio of 9 : 3 : 3 : 1.

It is now known that Mendel's second law is subject to exceptions (pp. 15–24). However, its simple operation must be discussed before these are considered, and this may be done most conveniently with the help of an example.

We will take for this purpose the character already used for illustrating Mendel's first law: the ability to taste phenyl-thio-urea, which is dominant to the inability to do so, and consider its segregation in relation to that of another character: the possession of bright red hair, which is unifactorial and recessive to non-red shades (see also p. 84).

The full red-haired condition, being recessive, must be homo-zygous, and is due to the operation of a pair of genes which may

GENETICS FOR MEDICAL STUDENTS

be denoted *rr*. The dominant non-red-haired state being represented as *RR* or *Rr*. Attention should be directed to this nomenclature. It is a convention to employ the same letter for allelomorphic genes, using the capital for that producing dominant effects and the small letter for that producing recessive ones. The initial letter of one of the characters is selected when appropriate. By this means the relation between allelomorphs is apparent when symbols alone are employed.

The double homozygous dominant (*TTRR*) is a taster with non-red hair. Men and women of this kind produce gametes possessing one member of each allelomorphic pair. They must therefore be all alike, carrying both *T* and *R*. Similarly, the double recessive (*ttrr*), which is a non-taster with red hair, produces gametes all equipped with *t* and *r*. On a marriage between individuals of these two types (the P1 generation) an F1 generation is produced which must consist of a single type; double heterozygotes of the constitution *TtRr* and these, manifesting the two dominant states, will be tasters with non-red hair.

Consider a marriage in which the partners are similar to this F1 generation. Both will carry *T* and *t* respectively upon one pair of homologous chromosomes, and *R* and *r* upon another pair, in every cell of the body. During meiosis the pairs of chromosomes will separate from one another into different gametes at random, in the sense that there is no tendency for the paternally and maternally derived material of each pair, the sections containing the spindle attachment, for example, to pass into the same gametes together. Consequently, the chances are equal that *T* becomes included in a gamete with *R* or with *r*, similarly for the equally numerous *t* genes. Thus four kinds of gametes, carrying *TR*, *Tr*, *tR* and *tr*, are formed in equal numbers by each individual.

These gametic types can be combined in sixteen possible ways, illustrated in Fig. 4. But they give rise only to four distinct classes, owing to dominance. It will be observed that nine of the recombinations contain at least one dominant member of both pairs of allelomorphs, and so are tasters with non-red hair, three possess *tt* and at least one *R* and are non-tasters with non-red hair, three possess *rr* and at least one *T*, being tasters and red-haired, while

12

one is of the double recessive type *ttrr*, a non-taster with red hair. These four groups arise in a ratio 9:3:3:1, which represents their frequency in such an F2 generation.

Only one true-breeding member of each of the four classes is included in a family of this kind (*TTRR*, *ttRR*, *TTrr* and *ttrr*). Consequently the proportion of homozygotes becomes progressively less in the larger of them. An inspection of Fig. 4 also shows the form of segregation which would have taken place in the absence of dominance. All the genetically distinct classes would then have been separable. There are nine of these,[1] segregating in a ratio of 1:1:2:2:4:2:2:1:1.

It is important to examine the situation arising from a back-cross involving two independently assorting allelomorphs. We have seen that the double heterozygote (*TtRr*), being a taster with non-red hair, must produce four types of gametes (*TR, Tr, tR, tr*) in equal numbers (p. 12). Also that the non-tasters with red hair, who are double recessives (*ttrr*), can produce one type only (*tr*). Clearly a marriage between two such individuals must yield all four classes of offspring, tasters with non-red hair and with red hair, non-tasters with non-red and with red hair. Since this is a back-cross, constituting an R2 generation, these will segregate in equality (pp. 7–8). This result is illustrated in the lowest horizontal line of Fig. 4.

We have now analysed the working of Mendel's second law in both of the fundamental genetic situations: those provided by the F2 generation and the back-cross. For this purpose we used the simplest instance, that in which two pairs of allelomorphs only are segregating. However, this second law indicates the ratios to be expected when three or higher numbers of allelomorphs are involved. Thus the 'tri-hybrid' ratio, that arising from a union in which both parents are heterozygous for three pairs of allelomorphs, leads, with dominance, to the combination of three 3:1 ratios independently: that is to say, to a ratio of 27:9:9:9:3:3:3:1. There are now twenty-seven genetically distinct classes and, without dominance, each of these would produce a separate effect. A mating between the triple heterozygote, manifesting all three

[1] They are *TTRR, TTrr, TtRR, TTRr, TtRr, ttRr, Ttrr, ttRR, ttrr*.

dominant characters, and the triple recessive will yield eight classes. These will appear in equality, for this is a back-cross in respect of all three factors.

F 1
gametes

		TR	Tr	tR	tr
F 1 gametes	TR	TTRR taster, non–red hair	TTRr taster, non–red hair	TtRR taster, non–red hair	TtRr taster, non–red hair
	Tr	TTRr taster, non–red hair	TTrr taster, red hair	TtRr taster, non–red hair	Ttrr taster, red hair
	tR	TtRR taster, non–red hair	TtRr taster, non–red hair	ttRR non–taster, non–red hair	ttRr non–taster, non–red hair
	tr	TtRr taster, non–red hair	Ttrr taster, red hair	ttRr non–taster, non–red hair	ttrr non–taster, red hair

FIG. 4. Segregation among the offspring of two double heterozygotes. The characters are the ability to taste phenyl-thio-urea, dominant to the inability to do so, and the possession of non-red hair which is dominant to the red-haired condition. Each heterozygote manifests the two dominant types, and carries T and t in one homologous pair of chromosomes and R and r in another. They therefore produce four classes of gametes in equality. These can be combined in 16 possible ways, producing a generation composed of: 9 tasters with non-red hair, 3 tasters with red hair, 3 non-tasters with non-red hair, and 1 non-taster with red hair.

In conclusion, it must be noticed that the operation of Mendel's second law provides an opportunity for genes which have arisen separately to be brought together. This is obviously of great value, since it allows advantageous characters possessed by different stocks or races to be combined in the same individual.

I.2 LINKAGE AND CROSSING-OVER

As already explained, it is now established that the genes are carried in the chromosomes (pp. 2–3, 207). This fact sets a limit to the operation of truly independent assortment, so that Mendel's second law is not invariably applicable.

The number of allelomorphs possessed by the human species is of course unknown. General considerations, depending on the possible size of the molecules constituting the genes, suggest that the total must amount to many thousands. There are twenty-three pairs of chomosomes in man, so it is obvious that each of them must carry a large number of genes. These cannot assort independently, as Mendel thought they did, since they travel in the same vehicle and consequently may be expected to reach the same destination. There is, of course, no barrier to the independent assortment of allelomorphs carried in different pairs of chromosomes, and this is the situation which we have so far analysed. Exceptions to Mendel's second law, of the kind just indicated, constitute 'linkage'. This should be defined in the following terms:

Linkage is the tendency for two or more pairs of allelomorphs to assort together, instead of obeying Mendel's second law of independent assortment, because they are carried in the same pair of chromosomes.

It will be convenient at the outset to illustrate linkage by means of a purely hypothetical example. Consider two heterozygous allelomorphs *Aa* and *Bb*, which are carried in the same pair of chromosomes. We will first suppose that *A* and *B* lie together in one homologous pair of chromosomes and *a* and *b* in the other. Consequently the genetic constitution of such a 'dihybrid' individual may be represented *AB/ab*.[1] It may be regarded as the F1 generation of a cross whose parents, of the P1 generation, we shall study subsequently. At meiosis the homologous chromosome-

---

[1] Genes within the same chromosome will be represented together on one side of a line, thus *AB/*. This represents the condition when no homologous partner exists, as in the gametes, the other side of the line being then blank. When both homologous chromosomes are present, the line will separate the allelomorphs carried respectively in them, thus *AB/ab*. I owe this excellent notation to Professor R. A. Fisher.

pairs separate, carrying $AB/$ together into one half of the gametes and $ab/$ together into the other half. Now double recessive individuals ab/ab can give rise to one kind of gamete only, bearing $ab/$. On mating these two types, a back-cross generation is produced consisting of two classes only, AB/ab and ab/ab, instead of the four classes which should arise with independent assortment (compare the example of this on p. 13).

Suppose the genes had been differently arranged in the double heterozygote, A and b being carried on one chromosome and a and B on the other, to give the constitution Ab/aB. On mating with the double recessive ab/ab, which produces $ab/$gametes only, two classes instead of the expected four will again arise; but this time they are Ab/ab and aB/ab.

These facts require brief consideration. In the first instance, the genes A and B were carried together in one chromosome of the double heterozygote and a and b were carried in the other. That is to say, the two controlling dominant characters were linked in one chromosome and the two controlling recessives in the other. On crossing with the double recessive, that same association is preserved, for only the double dominant and double recessive types appear. In the second instance, genes for one dominant and one recessive type were carried together in each homologous chromosome. This condition was preserved also in the segregating back-cross family, which contains two classes of individuals only, possessing one dominant and one recessive character each. The situation presented by the first example, in which the double dominants and double recessives tend to remain together, is called *Coupling* and the genes determining them are said to be in the *cis* position. That in which the two dominant characters tend to appear in different individuals, so that dominants and recessives seem to be repelled from one another, is *Repulsion*, and the genes controlling them are in what is known as the *trans* position. This is represented by the second instance.

Yet the parents of the coupling and of the repulsion phase were exactly the same in appearance. The double recessive was of identical constitution in each, and may therefore be disregarded. The double heterozygotes would manifest the two dominant conditions in both examples, though the distribution of the genes was different in them. Their homologous chromosomes contained A and B, a and b in coupling, but A and b, a and B in repulsion. This is due to the fact that these double heterozygous F_1 individuals had parents of distinct types. These are the grandparents of the final back-cross generation, and may be designated P_1. It is simplest to consider the situation in which such grand-parental forms are homozygous. In the coupling phase, they would be AB/AB and ab/ab respectively. In repulsion they would be Ab/Ab and aB/aB. That is to say, they are visibly distinct and, since we assume dominance, they are respectively of the same two classes as arise among their grandchildren.

It is clear, therefore, that in order to study linkage fully it is necessary to possess information on three generations. For the parents of a back-cross segregating for coupling or for repulsion between a given pair of allelomorphs will appear the same. It is only when we examine their grandparents (P_1) in the double heterozygous line that we again find a distinction between them. Linkage therefore involves no association between particular genes as such; there is no tendency, for example, for dominants or recessives to segregate with one another. Those genes which go into a cross together tend to come out of it together if they are carried in the same pair of chromosomes.

The theoretical examples of linkage so far considered are somewhat idealized, since they are instances of total linkage which, though known (in the male of the fruit-fly *Drosophila*, for instance), is very uncommon. They do, however, illustrate in a particularly simple way the manner in which the chromosome mechanism can limit the operation of Mendel's second law. Yet the facts of cytology will suggest that linkage can rarely be complete. The interchange of sections of material between chromatids derived from homologous chromosomes, during the prophase of the first meiosis, must carry blocks of genes with them. Such a transference, being an interchange of material without an interchange of partners, is demonstrated cytologically by a chiasma. Genetically it produces 'crossing-over', in which genes are transferred from one homologous chromosome to the other; for the prophase chromatids (pp. 206–9) become the daughter chromosomes of the second meiotic telophase. Consequently, linked genes are not irrevocably associated together, but can be combined in fresh ways, as can those carried in different chromosome-pairs.

Crossing-over may be defined as an interchange of linked genes consequent upon a reciprocal transference of blocks of material between chromatids derived from homologous chromosomes.

It will be well first to study crossing-over with the help of an example, and then to consider the general principles which it involves. Unfortunately, very little is yet known of this important phenomenon in man, except in relation to the sex-determination mechanism (to be explained in Chapter 2). Therefore it will be convenient first to examine its operation in another animal; attention is drawn to human instances of crossing-over during the course of this book.

In rabbits a gene Y produces an enzyme which enables the animal to oxidize any xanthophyll which it eats. The recessive condition yy prevents the formation of this enzyme, so that the fat becomes yellow in animals supplied with green food.[1] If this is not provided in the diet, the two conditions cannot be distinguished. Furthermore, albinism is recessive to normal coloration, and is due to the operation of a pair of genes cc. These are carried in the same pair of chromosomes as those determining yellow fat.

A homozygous normal rabbit, CY/CY, will produce gametes carrying $CY/$, while an albino with yellow fat, cy/cy, will produce gametes carrying $cy/$. If these animals, which constitute the P1 generation, be mated, all the F1 offspring will be normal in appearance but double heterozygotes of the *cis* type, CY/cy. They will form gametes in which the two linked genes generally remain together, carrying $CY/$ and $cy/$ in equality. However, crossing-over occurs between them approximately 14·4 times in 100, so that 7·2 per cent of the gametes becomes carriers of $Cy/$ while 7·2 per cent carry $cY/$. If such F1 animals be mated with albinos with yellow fat, being the double recessive type cy/cy,

[1] The character can easily be detected in living animals by snipping off a little skin between the shoulder-blades, so as to expose the subcutaneous fat.

which produces $cy/$ gametes only, a back-cross generation is obtained. This is composed of 42·8 per cent of normal offspring, CY/cy, and 42·8 per cent of albinos with yellow fat, cy/cy. Thus the two grandparental types have reappeared in equality. However, 7·2 per cent of these back-cross animals are normally coloured, but with yellow fat, Cy/cy, and 7·2 per cent are albinos with white fat, cY/cy. These two groups, which represent a transference of the P1 characters, are called the *recombination classes*. Together they measure the percentage-frequency of crossing-over. This is known as the *cross-over value* (C.O.V.), and is obtained by adding together the two recombination classes and expressing them as a percentage of the total number of offspring. In this instance the C.O.V. = 14·4 per cent.

The error involved may be assessed from the standard error of a percentage (p); that is to say, $\sqrt{p(100 - p)/n}$, where n equals the total number of individuals in all classes. Thus if in the rabbit example just cited 200 offspring had been accumulated by adding different litters and 34 of them belonged to the recombination classes, $\sqrt{17} \times 83/200 = 2\cdot66$. Our *estimate* of the C.O.V. would therefore be $17 \pm 2\cdot66$. This does not depart significantly from the accepted value of 14·4, which is based upon many results, since it falls within the range $17 \pm (2\cdot66 \times 2)$.

The example just described is one of Coupling. Repulsion could have been demonstrated by using homozygous fully coloured animals with yellow fat, Cy/Cy, and homozygous albinos with white fat, cY/cY, for the P1 generation. An F1 generation of normal, but double heterozygous, rabbits would have been formed as before, but with genes in the *trans* arrangement, Cy/cY. When mated to double recessives, albinos with yellow fat, cy/cy, a back-cross would be produced. This would consist, once more, of the two grandparental types in equality, each comprising approximately 42·8 per cent of the offspring (fully coloured animals with yellow fat, Cy/cy, and albinos with white fat, cY/cy), and the two recombination classes (normals, CY/cy, and albinos with yellow fat, cy/cy) also in equality, comprising 7·2 per cent of the offspring each. This is repulsion, with a deficit of the double dominants and double recessives. The same groups appear as in the previous example, but their frequencies are reversed. However, these remain approximately as they were in value, and the amount of crossing-over is still measured as 14·4 per cent by adding together the two recombination classes, though these are not the same as before. Once more we see that there is no association between the allelomorphs except that of their position on the chromosomes.

Since chiasma-formation involves two out of the four chromatids, its occurrence 100 per cent of times between two loci produces 50 per cent of crossing-over. This, which is normally the maximum value, gives the same effect as independent assortment which, therefore, occurs more often than might be expected in view of linkage (p. 15).

All the genes carried on a particular chromosome are of course linked with one another. It should therefore be possible to assign the linked genes of any organism to a series of groups. If the chromosome theory of heredity is valid, the number of such 'linkage groups' should equal the number of pairs of homologous chromosomes in the species. This proposition has been proved correct in all instances in which the genetics are sufficiently known to permit of testing it. Furthermore, when the pairs of homologous chromosomes differ from one another considerably in length, the size of the linkage groups varies roughly in proportion.

An important relationship exists between the cross-over values of three or more linked genes. If we consider three linked genes A, B and C, the cross-over value between A and C equals either the sum of the cross-over values between AB and BC, or the difference between them. This condition can only be satisfied by a linear relationship, and it demonstrates that the genes are carried in a linear series along the chromosomes. Moreover, it allows the correct order of the genes to be ascertained. Suppose the cross-over value between A and B is 4, and that between B and C is 1. Then the cross-over values between A and C will be either 5 or 3. If it be the sum (5), then the correct order of the genes is ABC; if the difference (3), the correct order is ACB.

The nearer together any two genes lie on a chromosome, the smaller is the chance that a break and interchange of material will occur between them. Therefore the cross-over values provide a relative measure of distance. In our last instance, gene B lies four times as far from A as it does from C.

This proposition is not true for a considerable length of chromosome, along which double, or even multiple, crossing-over may take place, but it becomes so for small cross-over values, say under 8 per cent. This is not due merely to the low random frequency of a chiasma forming twice along so small a distance, but one interchange tends to protect the chromosome from another coincident one in its immediate neighbourhood. That effect, known as *Interference*, is due to either of two causes: (1) to a genuine reduction of double cross-over frequency below the calculated value where the two occur in close proximity, since the chromosomes to some extent resist twisting; and (2) to a non-random association between the chromatids that are successively involved in the interchange. This latter point needs brief consideration.

When two chiasmata form between a given pair of chromatids, the interchange arising from one is annulled by the other if occurring in the same pair ('reciprocal' chiasmata). Thus the resulting cross-over individuals are scored in the non-cross-over class. If, however, the two chiasmata involve two different pairs of chromatids, the 'complementary' type (the two 'inner' and the two 'outer', respectively, in Fig. 12, p. 206), all four of them carry a single cross-over. The 'diagonal' arrangement in which one chromatid contains two chiasmata, two contain one, while the fourth is not affected, leads to the formation of a tetrad two members of which will be scored as cross-overs and two as non-cross-overs, though the latter for different reasons. Provided that, in all, the three types occur with random frequency, they ensure that 100 per cent of double crossing-over between two points on a chromosome leads, as with single crossing-over, to a cross-over value of fifty between them.

The result of Interference can to some extent be allowed for by a simple formula due to Kosambi. If x be the required distance apart of two genes, and y the observed cross-over value between them $\div$ 100, then

$$x = \tfrac{1}{4} \log_e \frac{1 + 2y}{1 - 2y}$$

Thus if the observed C.O.V. between vestigial and rex (in the mouse) be 27, $x = \frac{1}{4} \log_e (1 + 0.54 \div 1 - 0.54) = 0.302$, so that these loci should be placed 30·2 units apart.

It will now be evident that chromosome maps can be constructed from the data supplied by crossing-over. For the genes can be assigned to a number of groups each representing those situated on the same chromosome, while within these they can be placed in their correct order and at their relative distances apart. Naturally the latter step cannot be taken at all precisely unless chiasma-frequencies are constant along the length of a chromosome. However, they are far from being so, thus introducing an error here.

The fundamental method of detecting autosomal linkage and calculating cross-over values given in this Section is evidently inapplicable to Man since it involves experimental breeding. Such information can, however, be obtained by means of an analysis of human pedigrees. A simple and excellent technique for doing so was provided by Race and Sanger in 1962. It can be obtained from their 1968 edition and is also described by Clarke (1964, pp. 97-9). This method cannot usefully be shortened and would need to be reproduced as it stands, with its Table and formulae.

However, the required references having been given, it does not seem necessary to supply the calculations again here since, though a knowledge of the general theory of linkage and crossing-over is essential for understanding human genetics, the detailed analysis of these situations in particular instances is more likely to attract the attention of research workers than of medical students and practitioners. Yet there may well be some in general practice who wish to contribute to our knowledge of these important subjects (pp. 182-3). For this purpose, it will be seen from Race and Sanger, or from Clarke (*l.c.*), that they should keep a careful watch among the families they encounter for parents of the double back-cross type in which the two segregations are taking place among their children; for this situation provides particularly simple and informative information on linkage values. That is to say, by way of illustration, parents who can be scored for two unifactorial characters: such as (blood group A, non-tylotics) X (blood group

O, with tylosis). It is obvious that the occurrence of segregation among the children, in itself required for the analysis, demonstrates that the phenotypically dominant forms are heterozygotes in the parents.

I.3 MULTIPLE ALLELOMORPHS

The position of a gene upon a chromosome is known as its *locus*. The nearer the loci are together the closer will be the linkage between the genes. Those which lie exactly opposite one another are the allelomorphs. Thus we reach a definition of allelomorphism:

Allelomorphs (or alleles) are genes which lie at identical loci in homologous chromosomes. They control contrasted characters (p. 38), and segregate from one another according to the first law of Mendel.

It is sometimes possible for a particular gene to exist in more than two phases, so giving rise to a series of *multiple allelomorphs*. It will be apparent, however, that not more than two members of such a series can coexist in the same diploid individual. Multiple allelomorphs are of much importance in medical genetics, as will be found in discussing the blood groups. At this point they can, however, be more effectively illustrated with reference to other mammals.

Multiple allelomorphs usually control a given set of characters quantitatively, so that they can be arranged in a series representing the degree to which their effects depart from the normal. In these circumstances, the normal condition is usually dominant to the others, but when two of the mutant genes are brought together they produce an intermediate result (p. 24). It will be recalled that genes at the same locus are represented by the same letter, using the capital for that giving rise to the dominant characters. The fact that a series of genes occurs at the same locus is indicated by giving the same letter to all of them and distinguishing the members by a suffix. The normal allelomorph is represented by the capital without a suffix. Those recessive to it have small letters, the end term of the series being again represented without a suffix. Should genes dominant to the normal be included, they are indicated by a capital with a suffix. Thus the amount of black pigment in the hairs of the house mouse is controlled by the following series of multiple allelomorphs: A^y (yellow), A^W (grey with white belly), A (the normal uniform grey), a^t (black and tan), a (black). Here the effect of the last two genes is recessive to that of the normal (A), and that of A^W is dominant to it. A^y is lethal when homozygous.

When animals possessing different recessive characters are crossed, the offspring are of the normal type, provided that the genes are not at the same locus; for each carries the dominant allelomorph of the other. This is, of course, true even if the effects of the genes are superficially much alike. For example, the recessive pink-eyed condition in the guinea-pig produces, in addition to the change in eye-colour, a reduction in all the pigments of the coat except yellow. It is due to a pair of genes pp. Red-eyed dilution, $c^r c^r$, has similar effects except that the influence on coat colour is more particularly a reduction in the yellow constituent. On crossing these two types, $ppCC$ and PPc^rc^r, the offspring have the constitution $PpCc^r$. These are ordinary brown dark-eyed guinea-pigs. On the other hand, the recessive white guinea-pig with red eyes (strictly called the Himalayan type) is due to the operation of the genes $c^h c^h$, which are multiple allelomorphs of C, c^r, and of two others. Consequently, on mating a red-eyed dilute animal, $c^r c^r$, and a Himalayan, $c^h c^h$, the normal condition is not restored among the offspring since, being allelomorphs, one gene does not bring in the dominant allelomorph of the other. The F1 constitution is $c^r c^h$, and such individuals are intermediate between the two parental types. This result is diagnostic of the behaviour of multiple allelomorphs. It will further be evident that all the members of such a series have the same cross-over value with the other genes in their chromosomes.

I.4 THE ANALYSIS OF HEREDITY

Having now examined the 'Laws' of Mendel in the light of the chromosome basis which controls them and modifies his view of independent assortment, it is necessary to consider their place

MENDELIAN HEREDITY

among the general principles of heredity. Seven of these are fundamental.

1. It is a logical necessity that *units* of some kind, responsible in some way for the characteristics of the organism, must be passed from parent to offspring if heredity has any physical basis at all. Their chemical structure, the mechanism of their transmission and the degree of their permanence are not relevant as far as that essential concept is concerned. We can simplify this analysis from the outset by referring to these units as 'genes'; the name given to them by Johannsen in 1909, though he applied it strictly to the Mendelian situation.

2. *The two sexes contribute in approximate equality to the heredity of their offspring.* It is true that sex-linkage (pp. 47–58) constitutes a partial exception to that statement, but it does so to a very small extent (p. 57) and only because a number of genes are included within the super-gene (pp. 28–32) which actually determines sex. No exception however, is provided by sex-controlled inheritance (see pp. 81–2).

A proof of this general proposition is provided by a well-known mathematical technique, that of *correlation*, which assesses the extent to which two variables are associated. This is expressed by a number, the *correlation coefficient*, which is fractional and is capable of taking any value from 0, when there is no association between them, to 1·0 when the association is a perfect one (such as that between weight and diameter in a series of solid spheres of the same material but of differing sizes). A simple method of calculating it is given by Clarke (1964, pp. 328–33). The correlation coefficient can be employed to quantify the degree to which any measurable quality is independent as between parent and offspring, whether influenced much or little by the environment. For example, when studying a group of families, it can be asked to what extent, if at all, the taller fathers produce the taller sons (or daughters); similarly, for the same group of sons (or daughters) and their mothers. The correlation coefficient demonstrates that in each comparison there is an association in stature between the two generations and, moreover, that the contributions of either parent to the height of their children is effectively equal (though the

25

respective mean heights will of course be different; being greater for the fathers than the mothers, and for the sons than the daughters); for it can be shown that any observed difference between two such correlations is not mathematically significant (see Bulmer, 1970; with references).

Two essential conclusions can be drawn from this result.

(*a*) The hereditary units are, not entirely but to an overwhelming extent, carried by the nucleus of the cells. For while the egg contains thousands of times as much cytoplasm as the sperm, the nuclear material is transferred equally by the two gametes.

(*b*) Since the two sexes contribute in equality to the heredity of their offspring, the genes must be present in pairs, the members of which are derived respectively from the two parents. Indeed nothing simpler than the presence in the somatic cells of two corresponding members of each type of gene is possible in view of the equality just mentioned; a consideration true also of the chromosomes. Thus the existence of paired alleles is a necessary deduction from the information supplied by correlation. A slight adjustment to this statement must be made in respect of polyploidy. It will be inserted where that situation is briefly mentioned (pp. 63-4).

3. *Hereditary units (genes) do not contaminate one another when brought together into the same individual, or indeed into the same cell.* Thus they are 'particulate', that is to say, they are not subject to blending, which was assumed by most other theories of heredity. It was primarily Darwin's belief in 'blending inheritance' which was responsible for his relatively few though fundamental mistakes, including his acceptance of certain Lamarckian concepts.

A general test of this second proposition is also available. Let us compare any measurable quality in three consecutive generations, the third being strictly an F2; that is to say, obtained by brother and sister mating in F1. Thus mutation apart, see proposition 4, the grandchildren of the original pairing possess no genes not present among the children.

With blending inheritance, the F2 generation will be less variable than the F1, for blending leads to uniformity; that, indeed, is what we mean by blending. But if the hereditary units retain their identity and can be reassorted intact in new combinations, then the F2 generation will actually be more variable than the F1, and this is what we find: it is the antithesis of blending. A similar comparison can be obtained, though less easily, from the inbreeding of near relatives contrasted with normal out-crossing; one that is applicable to man.

It will be noticed that we are here concerned with the absence of blending in the hereditary material. The blending of *characters* is, on the other hand, frequent but is irrelevant to the present discussion which is concerned with the behaviour of the genes.

Any theory of heredity must be judged by its impact upon evolution, and from this point of view it must perform two apparently opposed functions: it must supply great heritable diversity, to provide the variability upon which selection can operate; also great heritable stability, to preserve the desirable qualities and combinations of qualities to ensure the correct working of the body and its adjustment to the environment. There is a paradox here. It is resolved by keeping the hereditary units themselves very constant but combining them in a potentially infinite number of ways, though by a system in which the association of certain genes can be maintained if necessary (pp. 28–33).

4. One step which makes this possible is due to the second proposition already quoted, that the genes do not contaminate one another when brought together; but that is not enough. In order to achieve genetic permanence, the hereditary units must themselves be intrinsically stable. That is to say, the chemical copying process undertaken when they divide must be very exact or the mistakes involved in it concurrently repaired: a statement which may be translated into genetic terminology by saying that mutation (a change in a unit of heredity, p. 62) must be very rare. So much is this true that if one individual in 80,000 be a mutant for a particular gene, we regard this as a very high mutation-rate; one in a million is probably more normal though, for practical reasons, we can measure only the highest frequencies of occurrences so exceptional (see Chapter 3).

We have recourse to another general test at this point, one due to R. A. Fisher (1930a, p. 18) though his argument is from a different point of view, for he uses it to exclude blending in the hereditary material. It can, however, equally well be employed to demonstrate the rarity of mutation. 'Pure-lines' (a group of individuals approximately homozygous for all pairs of allelomorphs) seem to be attainable whenever the requisite mating-system, self-fertilization, exists. For on selfing, all the pairs of alleles that are homozygous remain so and half of those that are heterozygous become homozygous at each generation. Yet as Fisher points out, a pure-line could never be approached unless mutation were very rare, otherwise its occurrence would oppose effectively the decay in variability due to selfing. Moreover, it may be added that selection experiments can establish the existence of a pure-line. They do so since they become unavailing as that condition is approximately reached because the necessary genetic variability for them to work upon is then lacking.

5. and 6. *These two propositions are represented by Mendel's two laws.*

7. *The evolution of super-genes.* We have seen that the chromosomes provide the physical basis for segregation as well as for independent assortment, subject to linkage; also for crossing-over. But they do more than that for, in addition, they supply a situation upon which selection can act to bring and hold together co-adapted genes so that they can segregate as a group or 'super-gene'. That aspect of their working has been so little regarded that it requires more detailed treatment here. It is therefore considered in the following section.

SUPER-GENES

It is evident that each of the complex adaptations and adjustments of the body, whether morphological or physiological, must generally require the co-operation of several or many genes. These will normally be kept homozygous if they are responsible for some quality that is constantly required in all individuals of a species: the correct development of the heart for instance. In these circumstances, such genes will produce only rare mutant forms which will nearly always be eliminated by selection, consequently they can be scattered among the chomosomes. On the other hand, we shall find in the chapters on polymorphism that appropriate and different 'co-adapted' genes have often to be held together so that they and their alternative alleles can respectively segregate from one another in a group. Yet how can this be done if they assort independently when on different chromosomes and are separated by crossing-over when on the same one?

That difficulty is overcome by building them into a closely linked series. When its members are so seldom broken apart by chiasma-formation that they act effectively as a single unit, they are known as a *super-gene* (Darlington and Mather, 1949). Its components will of course occasionally be separated by crossing-over, producing a different combination potentially as stable as the co-adapted one from which it was derived; but if this new arrangement is ill balanced it can then be eliminated by selection in the same way as can a disadvantageous mutant. The mechanism involved in the construction of super-genes must now briefly be outlined.

If unlinked major genes interact advantageously, structural interchanges (p. 65) bringing them on to the same chromosome will be favoured. These are not unreasonably rare. Thus Jacobs *et al.* (1971) found 4 such interchanges among 2,538 men, in Scottish Penal Institutions, whose chromosomes they examined. Subsequently, selection will reduce crossing-over between such genes, as it will if they chance to be linked initially; because as close linkage develops, so a smaller selection-pressure will suffice to maintain their more advantageous combinations (Ford, 1971, p. 112). That result can be achieved, in the first place, by diminishing the chiasma-frequency between them, for the positions of chiasmata are under genetic control (Darlington, 1956), or even by moving them nearer together by means of chromosome reconstructions. Indeed genetic variation in cross-over values is well known. For instance, in the snail *Arianta arbustorum*, the cross-over value between the genes for banding and ground colour is normally 1 per cent, but stocks have been found in which it has changed to 20 per cent (Cain *et al.*, 1960).

However, the most effective method of building two or more linked co-adapted genes into a super-gene is provided by inversions (p. 65). These may be either 'short' (too short, that is to say, to form counter-turned loops) or 'long' when counter-turned loops can arise within them. In those circumstances, the inverted region forms a loop turned upon itself, lying within an open loop formed by its counterpart on the homologous chromosome; thus the genes in the inverted and non-inverted lengths take up positions opposite one another. I have elsewhere discussed the way in which these two types faciliate super-gene formation (Ford, 1971, pp. 110–16). It is unnecessary therefore to give a detailed account of that subject now. Here it need only be said that viable chiasma-formation can of course take place within all homozygous inversions since corresponding loci are then arranged in the same order on both homologous chromosomes, and therefore on both pairs of homologous chromatids when formed, whether in the inverted or non-inverted sections.

'Short' inversions present an almost complete barrier to effective interchange in structural heterozygotes, since in them corresponding loci no longer lie opposite one another. They may perhaps sometimes come to do so by torsion when, however, as Darlington says, crossing-over probably fails within the affected region.

'Long' inversions, in which counter-turned loops can arise, so allowing crossing-over within them, present a more complex picture owing to the diverse conditions involved: whether the chiasmata are extracentric or pericentric, and whether one or several are included in the inverted region. But they generally destroy the chromosome-pair, or act at least in a very disadvantageous way.

We need here only consider the simplest situation: inversions producing counter-turned loops and containing a single crossover. This when extracentric (not including the centromere) produces one acentric and one dicentric chromatid and the latter, being pulled in opposite directions by its two centromeres, results in a bridge whose breakage is irregular. The acentric, being without means of orientation, is eliminated, giving rise to a probably fatal loss of genetic material (deletion, p. 65) or if very small, to a dominant mutant with a characteristic syndrome.

Such loops forming inversions may also be pericentric (including the centromeres). These give rise to chromosomes lacking some genes and carrying an excess of others, so that the individuals which come to possess them are at least partly sterile. Thus while inversions are highly advantageous if they hold together co-adapted genes, there is strong selection against any cross-overs which break such groups apart.

Evidently the evolution of super-genes constitutes one of the great principles of genetics, for it represents the converse of Mendel's second law and provides a mechanism by which favourable genetic combinations can be preserved. This carries with it, as a corollary, the proposition that the genes do not necessarily and always occupy fixed positions, as they are shown to do in the *Drosophila*, and other, chromosome maps. On the contrary, their order and their cross-over values can be altered by selection, even experimentally in the laboratory. This fact has long been recognized by some geneticists but ignored by others, even though the effects of selection upon crossing-over have been effectively demonstrated (see Levene and Levene, 1954; Parsons, 1958).

It may be inquired, therefore, if the order of the genes on the chromosomes and their relative distances apart are as haphazard as they superficially appear to be from the chromosome maps. To put it another way, how is it that the positions of the genes are always approximately the same whenever mapping of a given species, or at least sub-species, is undertaken? The answer of course is that the positions of the genes are by no means haphazard, and that the stabilization of their loci is due to the fact that these are not randomized. No doubt large numbers of the 'normal alleles' lie in proximity because they are co-adapted while others will be carried with them in building up the chromosome construction which facilitates that situation.

It is unfortunate that, in general, cytology has not yet advanced sufficiently to demonstrate the small unpaired regions during meiosis which represent short inversions. Only when the polytene nuclei of Diptera could be studied was it found, with a surprise indeed surprising, that small inversions recognizable in heterozygotes are scattered widely along the greatly elongated chromosomes. Long inversions can, however, sometimes be detected in normal cytological material owing to the loops which characterize the structurally heterozygous segments; a situation which has been demonstrated in man, as first shown by Koller (1937). Jacobs *et al.* (1971) found one pericentric inversion of a medium-sized autosome among 2,538 men. In the same group, they also detected two metacentric Y-chromosomes, both familial and present in all the male relatives who could be studied. They considered that the condition resulted from pericentric inversions.

It will be evident that a considerable section of a chromosome may constitute a super-gene: as with the differential, or sex-determining region of the sex chromosomes; so also, in special circumstances, may a whole chromosome or even a chain of chromosomes. But that situation, like the super super-gene of the Evening Primrose *Oenothera lamarckiana*, Onagraceae, where the entire gene-complex acts effectively as a single switch-unit (Darlington, 1956, p. 38), is outside the scope of a work dealing with human genetics.

It may be remarked that super-gene formation holding together little groups of genes, must be partly responsible for the inheritance of the striking resemblances and characteristics which are so marked in some human families. Their effects will often pass unnoticed, but not when they control features giving rise to family likenesses. It is possible also that even a single major gene may sometimes influence a characteristic important from this point of view, such as the proportions of the bones of the face. Together with the super-genes already mentioned, it may contribute to the strong similarity sometimes seen between members of a family in the same and different generations or between an individual and a remote ancestor or collateral.

The account of genetics generally given in elementary textbooks is one-sided and misleading. For it concentrates upon heritable *variability* and ignores the equally important converse, that of heritable *stability*. Yet these two conditions totally opposed though they be, are both ensured by the particulate character of Mendelian inheritance. Thus one aspect of its action provides the diversity upon which selection can operate, the other ensures that advantageous genes and combinations of genes can be preserved as needed. The latter effect is precisely what no genetic system could promote were heritable diversity and the control of evolution cytoplasmic, or else the outcome of a high mutation-rate as in the Lamarckian concept.

CISTRONS

It is clear then that the genes, strung out as they are in linear order along the chromosomes, are not arranged at random as to function. For here and there they are built into co-adapted groups, the super-genes, each controlling a given set of characters. A linear organization occurs also at another and smaller scale in the chromosomes. For each gene itself consists of a number, up to 300, of mutational sites forming, collectively, a *cistron*. Such loci are occupied by genetic units which control the same set of characters. They are non-complementary in their functions instead of determining sets of distinct qualities, as the various major genes, including those within the same super-gene, generally do. Intra-cistronic mutants consequently behave as multiple alleles, producing intermediate effects when heterozygous. On the other hand, the major genes existing at different loci complement one another even when the effects by which we judge their presence are apparently similar. Thus in the Amphipod Crustacean *Gammarus chevreuxi*, mutants at the loci of two major genes each produce recessive red eye-colour: that phenotype having respectively the constitution *rr SS* or *RR ss*. Yet when crossed, the F1 generation is of the dominant black-eyed form *Rr Ss* since, as these genes are complementary, the eggs and sperms each transmit a normal allele of the opposite type.

Moreover, by definition, intra-cistronic mutants give rise to the '*cis-trans*' effect. That is to say, when evoking a recessive character, the double heterozygote is non-mutant in the *cis* arrangement, *AB/ab*, but mutant in the *trans* one, *Ab/aB*.

Intra-cistronic units have probably been built up by side-to-side multiplication (duplication), and by selection operating upon them over great periods of time. It is thus that many of the normal alleles will have been evolved. For some major genes, the process may well have taken longer than the evolution of man; for others, longer than that of the primates or even of the mammalia. The centromeres and nucleolar organizers on the chromosomes are believed to be such replications and they presumably go back to the origin of mitosis. Conversely, the linear evolution of the chromosomes as super-genes, is likely to be a more recent process related to the current needs of micro-evolution. However, both mechanisms may be pressed into service even in relatively recent adaptations among which units responsible for similar types of characters are required. Thus Clarke *et al.* (1968) have found that a super-gene controlling mimicry in the butterfly *Papilio memnon* is compounded of five major loci. They produce evidence to suggest that one of these evoked the original mimic pattern, that two were brought from other parts of the chromosomes, while two were pulled out of the same cistron by duplication.

The concept of cistrons is apt to present difficulties to those who have been accustomed to think of the genes as indivisible units. It may therefore be helpful to summarize a few facts about it.

Each gene consists of a number of sites at which mutation can take place, and these together constitute a cistron. When an 'ordinary' mutation occurs, this means that *one* of the mutational sites within a cistron has mutated.

The distinct genes, which complement each other, may or may not lie in very close proximity: the closest seem to give recombinations of about 1 in a 1,000; or of course they may be widely separated, even by the whole length of the chromosome. On the other hand allelism, represented by sites within the same cistron, is *always* associated with very close linkage (less than 0·5 per cent).

All the sites within one cistron (gene) control the same set of characters, being non-complementary. Some of them, at least, may do so in slightly different degrees, and changes at more than one of these therefore give rise to multiple alleles. It is reasonable, therefore, to deduce that these can originate in either of two ways: (1) as mutations at different sites within the same cistron; (2) as a given site within a cistron mutating at different times to produce slightly different effects. When a rare cross-over does take place between members of a multiple allelic series, the first type evidently has operated. Otherwise it is hardly possible to distinguish between these two alternatives.

The major genes may still be regarded as the fundamental entities of genetics for all ordinary purposes. Their subdivision into cistrons is important only when we consider the minute structure and evolution of the chromosomes; it will nevertheless be seen that on occasion it has relevance to medical genetics (with respect, in particular, to the Rhesus blood groups, p. 156).

The evolution of the genetic material as a balance system is a concept due to the genius of Darlington (1939; see the 2nd edition, 1958). That fundamental work should be consulted by all those who require information on this subject.

THE STRUCTURE OF THE GENES

The structure and action of the genes has been elucidated since this book was first published. These are indeed matters of great importance, but as they are not at present closely relevant to the needs of medical men, they will be briefly summarized only (for further information, see Beadle, 1960).

The genes consist of deoxyribonucleic acid (DNA). The nucleic acids, of which this is one, are made up of pentose sugar, phosphoric acid and nitrogen bases comprising pyrimidines and purines. From the pyrimidines are derived cytosine and thymine while here the purines exist as adenine and guanine. The first step in the organizing of DNA is the formation of a nucleotide comprising one molecule of a nitrogen base, one of pentose sugar and one of phosphoric acid. This polymerizes to contribute to a DNA molecule which, being repetitive, forms a chain. It is one of two, similar in type, which constitute DNA. The two chains pair in such a way that a purine base always partners a pyrimidine one: adenine with thymine, cytosine with guanine, being the only combinations that provide links of the correct lengths between the chains (Fig. 5). These base pairs can follow one another in any order, and there may be up to 10,000 of them in one molecule. The chains themselves consist of the phosphate and sugar, and the nitrogen bases on the inner sides are united by hydrogen bonds. Their arrangement follows an opposite order in the two chains, which twist round one another in a double helix (Fig. 6).

The genes replicate, and Watson and Crick have suggested that, when the chains separate, they will each be left with a series of bases needing partners, which are derived from the nucleotide stock in the nucleus: each double chain is then formed of one old and one new component. Differences in the sequence of the base

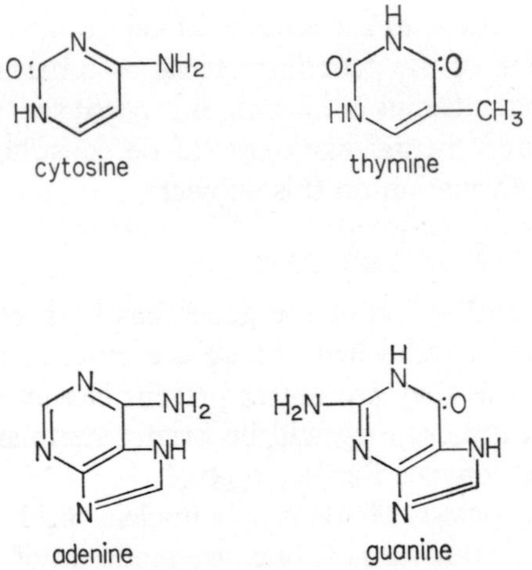

cytosine   thymine

adenine   guanine

FIG. 5.

pairs are responsible for the distinction between the genes, while mutation consists of a change in that sequence due to an alteration in the pairing of the bases. Those that involve the replacement of one purine, or else one pyrimidine, by another are known as 'transitions', while 'transversions' involve the replacement of a purine by a pyramidine or the reverse. The sequence of base pairs provides a code for the linear order of the amino-acids characteristic of different proteins.

The number of base pairs of DNA needed to determine a specific amino-acid is three. If the coding were effected by two only, there would be but 16 combinations of the four base types, which would not be enough to specify the 20 or more amino-acids. But the 64 combinations to be had from three base pairs are more than adequate to do so.

Instructions must be passed from the nucleus to the cytoplasm, where the proteins are made. This is achieved by another nucleic acid, RNA (ribonucleic acid), which has a very similar structure to DNA. Messenger RNA is formed against the DNA template with complementary base pairs. The amino-acids are then arranged in the order specified by the bases in the RNA.

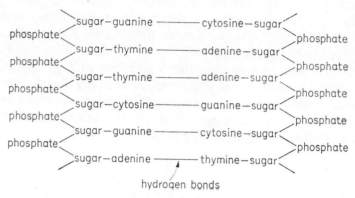

FIG. 6. The structure of DNA. (The two chains, united by hydrogen bonds, are twisted in a double helix.)

As long ago recognized (Ford, 1931), it seems that the volume of cytoplasm is at first too great to be controlled by the zygote nucleus, the proteins in which have, therefore, to be conditioned by the maternal genes. Only when, during successive cell divisions, the nuclear-cytoplasmic ratio returns to something near normal can its own messenger RNA be in effective command, and only then are the characters of the individual determined by its own nuclei. Up to that time, they are Mendelian, but imposed by the genotype of the female parent.

This situation was originally detected in hybrids from wide outcrosses in Echinoderms, and it has now been found in hybrid Salamanders (*Triton*). In addition, the result of such delayed inheritance may sometimes persist in adult organisms, as in Gastropod torsion; while instances of it are sometimes evident in the gametophyte growth of plants: that is to say, in heteromorphic incompatibility.

The number of forms in which this mechanism has been studied is few, partly because it has rarely attracted the attention of research workers, and partly owing to the difficulty of obtaining even the initial stages of development in suitable hybrids: those between species so remote in classification that they can be distinguished prior to early gastrulation. Yet the situation is a logical necessity, and any phenomenon which is known to occur in higher plants, Echinoderms, Molluscs and Amphibia must surely be a fundamental one (for further information and references, see Ford, 1971, pp. 207-8).

The chemical substitutions that a gene can undergo appear to be restricted, and it is this which enables us to say that the allelomorphs control contrasted characters. Thus we do not find a gene controlling, for example, eye-colour but not height mutating to one that controls height but not eye-colour.

CHAPTER 2

The Genetics of Sex

2.1 SEX-DETERMINATION

Sex is a quality which is inherited, and in a simple way. For the two contrasted characters involved, maleness and femaleness, with all their attributes, segregate sharply from one another in proportions which, in the adult (p. 58) are nearly equal. How complete this segregation may be is a matter of some dispute (pp. 46–7), but individuals obviously intermediate in structure between the two sexes, though known, are of great rarity. The mechanism which operates this system must briefly be considered.

The genes controlling sex are carried in a special pair of chromosomes, the 'sex-chromosomes'. The remaining chromosomes, collectively known as 'autosomes', are not directly concerned in actual sex-determination. In women, the two sex-chromosomes are similar to one another in shape and they do not differ considerably from the autosomes, except that they bear, among others, the genes determining sex. They are called 'X-chromosomes'. Men, on the other hand, possess an unlike pair of sex-chromosomes. One of these is an X-chromosome derived from the mother, while the other, received from the father, is an atypical chromosome carrying relatively few genes. This is the 'Y-chromosome'. Thus in the human species, with a total of 46 chromosomes (23 pairs), there are 44 autosomes (22 pairs) and one pair of sex-chromosomes, consisting of two X-chromosomes in women and one X- and one Y-chromosome in men.

During meiosis, the chromosome number is halved in such a way that each gamete acquires one member of every chromosome-pair. Consequently, in addition to the complement of 22 autosomes, all the eggs will possess a single X-chromosome, while half the sperms receive an X- and half receive a Y-chromosome. Thus the chances should be equal that any egg, necessarily carrying one X-chromosome, is fertilized by a Y-bearing sperm, so producing a male (XY), or by an X-bearing sperm, producing a female (XX) (see Fig. 7).

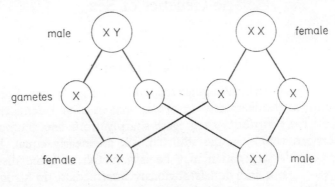

FIG. 7. Sex in *Drosophila*. This is determined by the number of X-chromosomes in the cells, one producing a male and two producing a female. This should ensure the production of the two sexes in equality in the next generation. Note that the sex-control is quantitative, since the X-chromosomes are interchanged between the sexes at successive generations.

The genetics of sex-determination were originally established by work on *Drosophila*. In that genus, and apparently widely in animals, it is clear that the Y-chromosome does not control sex, though it has other effects, while both autosomes and X-chromosomes contain numerous genes some tending towards female and others towards male development. The over-all effect of the autosomes is male determining and of the X-chromosomes female determining. The balance between them is so adjusted that the fixed dose of male-determinants provided by the autosomes outweighs that of female determinants in one X but is outweighed by that in two. A simple sex-switch on a quantitative basis is thus achieved (Fig. 7).

It is curious, however, that within the ambit of such a sex chromosome mechanism, generally though not universally consistent, the action of sex-determination varies rather widely. In the majority of organisms, including *Drosophila*, man and apparently all mammals, the sex with unlike (XY) chromosomes, known as the *heterogametic* sex, is the male; yet in some other groups, birds and Lepidoptera, it is the female and in these the male is *homogametic*: that is to say, XX. In some fish, Winge (1932) has shown that the sex-chromosomes carry one master-gene for sex. Moreover, the values of the other sex-determinants can be so varied by segregation and selection that even in a single species, *Platypoecilus*, the X-chromosome can be experimentally converted into Y and Y into X, while sex-chromosomes can be transformed into autosomes and the reverse. There are also groups, though not among the Vertebrates, in which Y is absent; the sex-control being then maintained on the basis of XO and XX. On the other hand, in the Amphibia the Y-chromosome is important in sex-control.

In view of these facts, it is not surprising to find, even in a relatively primitive placental mammal such as man, substantial differences from what may be regarded as the fundamental *Drosophila*-like sex-mechanism. Indeed it appears that the human situation is one in which the Y-chromosome is an important sex agent, carrying male-determining genes balanced against female determining ones in the autosomes. Thus Klinefelters (pp. 42-3), who are, chromosomally, women in which a Y chromosome is added (XXY), are predominantly masculine; while those who suffer from Turner's Syndrome, being chromosomally of the male type but lacking Y, and therefore XO, are physically infantile, and highly abnormal, girls. It has even been suggested that in the human species no genes for femaleness are carried in X, the balance being due solely to the autosomes acting in the presence or absence of the masculinizing Y. Since, however, Klinefelters are not normal men but are sterile (they hardly ever form sperm) and physically somewhat feminized, that point of view cannot be sustained, as indicated also by occasional XXXXY Klinefelters.

It should be mentioned that Lyon (1962) has proposed what is to some extent an alternative hypothesis. She suggests that only one X-chromosome of a pair is genetically active, it being immaterial in any cell in the same individual whether this be of paternal or maternal origin. There is indeed full evidence for a distinction between the two X-chromosomes in women, since thymidine, labelled by tritium, can be incorporated into both of them but the grains are taken up much earlier, indicating DNA synthesis, in one than in the other member of each pair. Lyon's view was based upon the situation in mice, in which chromosomally aberrant XO individuals are normal fertile females, which therefore do not require two X-chromosomes for their production; while sex-linked genes affecting coat colour give rise to a mosaic pattern in the heterozygotes. It was indeed owing to these two considerations that Lyon propounded this concept: but from what has already been said, it will be clear that it is a hazardous proceeding to equate the sex-determination of one group of animals with another. In this instance, involving Rodents and Primates, it does not seem justified, since we have evidence that the sex-control differs in them; for, as already mentioned, the human XO type, though female, is neither fertile nor otherwise normal. Moreover, Reed et al. (1963) have shown that the mosaic effect in sex-linkage is not applicable to man in the sex-linked blood group, in which it can be tested. Consequently, if Lyon's hypothesis holds, it does so for certain loci only. Lyon herself attempts to overcome the difficulty introduced by the Klinefelter and Turner conditions by suggesting that the inactivation of one X-chromosome in each female cell takes place only after the first sixteen days of gestation, a view for which there is some evidence, and that these two abnormal phenotypes are due to the action of both X-chromosomes in the earliest stages of development; alternatively, that only part of the human X-chromosome is inactivated.

Grüneberg (1967) made a detailed analysis of genes carried by the human X-chromosome in order to evaluate the Lyon hypothesis. He did not in any instance find support for it. It perhaps remains possible, however, that it may apply to certain loci (Clarke, 1969, p. 61).

Summarizing, the human situation involves feminizing genes carried principally in the autosomes and additionally in X. These are balanced against predominantly male-determining genes in Y.

The Klinefelter and Turner conditions, so far mentioned merely to illustrate sex-determination, must now be considered in more detail. They are reciprocal, being the result of non-disjunction of the sex chromosomes (p. 209). This may occur in the gametic formation of either sex. In the male it can give rise in equal numbers to XY and O sperms. At fertilization these contribute to form two types of zygote, XXY (Klinefelter) and XO (Turner Syndrome). In the female the corresponding process leads to the development of XX and O eggs which, on fertilization respectively by normal X- and Y-bearing sperms, produce four gametic types in equality: XXY, XO, YO and XXX. Here, in addition to the Klinefelter and Turner individuals, two other combinations arise; YO which die and XXX females. These latter differ little from normal women and may even be fertile, but they have actually been detected in a few instances only.

Klinefelters behave as males and often pass as such, although the breasts are enlarged, the face is smooth and the voice higher than normal. They possess the 'Barr body', reflecting the XX condition, in cells of the appropriate tissues (e.g. buccal mucous membrane). Such abnormal people are liable to mental deficiency, but many are of fair intelligence. Their marriage, which sometimes takes place, can be consummated, though they are almost always sterile since they hardly ever form sperms. As might be expected, Klinefelters frequently have psychological difficulties owing principally to their large breasts, which may give rise to comment especially in the armed forces, and lack of children when married. However, many must surely live out their lives unrecognized. Indeed it now seems likely that about one 'man' in 1,000 is a transformed woman of this kind. There is at least some evidence that Klinefelters are unduly subject to carcinoma of the breast (Harnden et al., 1971).

The converse situation, that of Turner's Syndrome, is far more serious. Those affected by it are girls who remain infantile. They are stunted and mentally defective. Not only are they sterile but they hardly ever menstruate. Moreover, they frequently suffer from coarctation of the aorta and other dangerous symptoms. If recognized in the first few years of life, their condition can be improved by administration of thyroid extract and testosterone.

Though, potentially, the occurrence of Klinefelters and of

Turner's Syndrome must be numerically equal, the latter class is much the rarer in the population since it frequently leads to early death. On the other hand, it is much more often detected, as advice is almost always sought on behalf of the patients owing to the absence of menstruation, while the Klinefelter state may pass unnoticed.

It has been found by Jacobs *et al.* (1965) and repeatedly confirmed by others, that the presence of an extra Y-chromosome in man (XYY) tends to be associated with increased stature and with aggressive criminal behaviour. In the Maximum Security Hospital for subnormal men where this was first detected, the mean height of 189 of the ordinary XY type was 67·0 inches; while that of eight men with an extra Y-chromosome was 73·1 inches, the difference being heavily significant. The situation is not restricted to those officially classed as mentally subnormal, for the high frequency of XYY individuals among tall men convicted of crimes of violence has now been established in English prisons, as well as in those of Australia and the U.S.A. (Jacobs *et al.*, 1971). It seems that an extra Y-chromosome is present in about one man in 512 in Indo-Europeans (McWhirter, 1970), and there is an undoubted tendency among them towards intractable and violent behaviour. We have here a good instance showing that the mentality of an individual is affected by his chromosome complement: and of course it is well known that intelligence is in general under genetic control (pp. 184–5).

It should, however, be noticed that in more recent work carried out in a number of Scottish penal and corrective institutions, Jacobs *et al.* (1971) found that the number of XYY men did not differ significantly from the frequency of that condition among the newborn though there is some evidence to the contrary in Canada. They point out that this may be attributable to the legal system and penal conditions in Scotland which differ from those in other countries, in which an association between this chromosome abnormality and criminal behaviour seems to be clearer. Thus the number of beds in institutions for male offenders in Scotland is 50 per cent greater relative to the population than in England.

While it is agreed that XYY men are exceptionally tall, doubt

44

has lately arisen about their previously reported aggressive behaviour which, it is held, may be due to a bias in ascertainment. Approximately one in 700 male babies are of this chromosome type, so the presupposition is that many are normal (Clarke, 1971a). This does not of course exclude the possibility that the *proportion* of those among them who are criminal or aggressive may be above the average. Indeed this is highly probable in view of the rather varied results that have been reported.

The embryological and physiological aspects of sex must now be related to the chromosome mechanism. The early embryo is capable of development either as a male or as a female, having a complete outfit of the rudiments required to build up one or the other type of sexual structure. An excess of male-determining genes causes the degeneration of the cortical, or ovarian, region of the gonad, and the development of its medullary, or testicular, element. The secretion of the male hormones from the interstitial cells of the testis ensures the growth of the male secondary sexual characters: vas deferens, penis and associated glands. It also causes the retention of mesonephric material such as the epididymis, which is carried backwards with the descending testes. The same hormones bring about the degeneration of the Mullerian ducts and other pronephric structures. On the other hand, an excess of female-determining genes brings about the reverse set of changes owing to the formation of ovaries and the secretion of the female hormones from their interstitial cells. Thus the secondary and accessary sexual characters of both sexes are under hormonal control, being especially stimulated to develop at the time of puberty.

In man, therefore, the chromosome outfit determines the type of gonad which shall develop, but only indirectly the secondary and accessary sexual characters through the intermediacy of the hormones from the testes or the ovaries. Yet this arrangement is by no means universal, for in many forms, the insects for example, all sexual characters are directly determined by the sex-genes. In such animals castration has no effect beyond sterilization, whereas the changes of puberty cannot take place in man if castration has been performed early in life. Thus the voice of eunuchs remains high-pitched and child-like.

It is by no means clear to what some of the human physical aberrations of sex are due. Yet it is important to notice that though the sex-genes decide whether the gonads shall develop as testes or as ovaries, many other genes will determine the amount of hormone produced and its timing, as well as the response of the tissues to it. Such genes, as well as those directly concerned in sex-determination, can, of course, mutate to other allelomorphs, which will allow of segregation in the usual way. Some of the combinations so produced may cause a weak or abnormal response either to the male or to the female hormones, giving rise to instincts inappropriate to the physical sex, perhaps combined with some slight tendency to approach the opposite sex in the proportions of the body or in the quality of the voice. Such are some at least of the homosexuals: though the psychological attributes of that condition are very liable also to be environmentally produced, but more easily in some constitutions than in others; as shown by the decisive work of Kallmann (1952) on homosexuality in identical, compared with fraternal, twins (pp. 186–7). It is important also to notice that the development of the full sexual activities is reached gradually, while genes exist which control the rate of processes in the body and the time of their onset (pp. 103–4); indeed the facts of inter-sexuality demonstrate that the action of the sex-genes is sometimes (and perhaps always) of this type. It is not surprising therefore to find that a phase of homosexual instinct is common in the adolescence even of sexually normal males and females. The length of time that such instincts persist varies, no doubt both genetically and environmentally. But they are frequently prolonged into adult life and become a fixation.

It is important that a physician should not give such homosexuals as he may have to advise the impression that their condition is the product of some shameful physical or mental state. On the contrary, such people need to realize that it is a frequent one, the result of ordinary genetic and, to some extent environmental, processes. As a result of the Sexual Offences Act 1967, homosexual practices between consenting adults are in general no longer unlawful in England and Wales; as they were previously between men, but not between women. When, however, they involve minors, or are committed in public or without the consent of one of the parties, they remain illegal, as indeed with heterosexuality. This Act does not apply to males in Scotland or Northern Ireland, and it is most desirable that the attitude to homosexuality underlying it, which is that widely adopted in other civilized countries, should at once be extended throughout the United Kingdom. In that way, the cruelty of criminal proceedings against homosexuals will largely be avoided, except when a juvenile partner is involved, and the blackmailer deprived of a ready weapon. We may hope also that the 'social blackmail' to which homosexuals are still to a considerable extent exposed, will be mitigated by the spread of knowledge relating to human development, physical and mental.

Gross abnormalities in the structure of the sexual organs, suggesting intermediacy, are extremely rare in the human species, and they appear almost always to be in the direction of an apparent feminization of males. But it should be noticed that the embryonic condition of the external genitalia is far more like that of the adult female than of the adult male. A mere retardation in male development may therefore give the impression of an approach towards the female type. Such an effect will usually be genetic, and the resulting abnormality is embryonic rather than hermaphrodite. No doubt, however, true sexual intermediates are occasionally produced.

2.2 SEX-LINKAGE

In the human species, the sex-chromosomes contain many more genes than those concerned with sex-determination. These affect the widest range of characters and bear no relation to sex, except that being carried in the same vehicle with the sex-genes they are distributed relative to them; whereas sex is not related to the segregation of the ordinary autosomal genes which have been considered in the previous chapter. It has been explained (pp. 15–22) that genes carried in the same chromosomes are said to be 'linked' because they assort together. Consequently, those in the sex-chromosomes are called 'sex-linked genes', and they give rise to 'sex-linked characters', owing to their association with sex.

The human X- and Y-chromosomes are represented diagrammatically in Fig. 8. The greater part of the X-chromosome, from *b* to *c* in the figure, is not homologous with Y, so that no crossing-over, and consequent interchange of material, can occur in this region. However, one short section of X is thought to be homologous with part of the much smaller Y-chromosome (*a* to *b* in

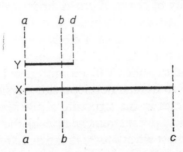

FIG. 8. The human sex-chromosomes. The large X-chromosomes and the small Y-chromosome each appear to possess a section (*a–b*) composed of homologous material. There is also a long section of the X-chromosomes (*b–c*) which is not homologous with material in Y, and a short section of the Y-chromosome (*b–d*) which is not homologous with material in X.

the figure), while a very short length of Y (from *b* to *d*) is not homologous with any material in X. Thus three types of sex-linkage may occur in man, due to genes carried: (1) in the non-homologous region of X (*b* to *c*), (2) in the homologous regions of X and Y (*b* to *a*) and (3) in the non-homologous region of Y (*b* to *d*). These must be considered in some detail.

In recent years it has seemed possible that the pairing of the X and Y chromosomes in the mammalia is abnormal, being terminal. This would not accord with the existence of partial sex-linkage (type 2 above) nor with those instances in which it seems to occur genetically. However, as Dr C. E. Ford points out, pairing with chiasma formation between short terminal segments of X and Y, followed by early terminalization, would give rise to a condition morphologically indistinguishable from terminal pairing and evidence for a situation of this kind is accumulating. He kindly allows me to say that in his opinion there are short homologous segments in the X- and Y-chromosomes of man, and of mammalia generally, between which pairing with chiasma formation occurs, though the present evidence for this falls short of actual proof.

(a) TOTAL SEX-LINKAGE IN THE X-CHROMOSOME

This is due to genes carried anywhere in X, except for the small region which appears to be homologous with part of Y. It involves the situation always intended when sex-linkage is mentioned, except when stated to the contrary.

A notable example of such a sex-linked character is provided by haemophilia. This is a disorder in which blood clotting is delayed. Blood oozes, sometimes for weeks, following even a slight injury, so producing severe anaemia. It is due to the absence of one of the clotting factors (No. VIII). Slight blows may give rise to severe bruising and extensive haemorrhages under the skin. In addition, effusions of blood into the joints may gradually produce stiffness and constant pain, to deaden which haemophiliacs sometimes become drug-addicts. Yet it is to be noticed that they stand operations quite well if previously transfused. However, such men usually die before the age of reproduction, but some live to be 20 or 25 and produce children. The condition has attained notoriety since it is rather widespread in the royal families of Europe.

Haemophilia is recessive and due to the action of a sex-linked gene. A woman, heterozygous for it is therefore unaffected, since she carries the haemophilia gene (h) in one X-chromosome, and its normal allelomorph (H) in the other. Consequent upon meiosis, half the eggs which she produces will contain h, and half H only. She will usually marry a normal man, who has H in his only X-chromosome and produces in equal numbers X-bearing and Y-bearing sperms. From such a union four types of offspring will arise in equal numbers, though they appear only as three recognizably distinct classes: normal sons and haemophiliac sons, and normal daughters. Of the latter, one half are homozygotes, but the other half are heterozygotes who are 'carriers' and can transmit the disease. That is to say, normals and haemophiliacs appear in a 3:1 ratio, but this is not distributed at random relative to sex as have been the ratios considered in the previous chapters. On the contrary, half the sons but none of the daughters suffer from the abnormality (Fig. 9).

On the rare occasions when a haemophiliac marries, his wife will usually be normal. All her eggs therefore carry H, while half her husband's sperm will contain a Y-chromosome and half an X necessarily carrying h. All the children will therefore be free

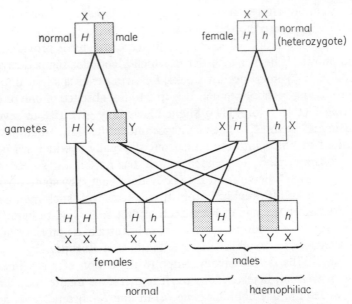

FIG. 9. Total sex-linkage in the X-chromosome: a marriage between a man with a dominant character and a heterozygous woman. The gene producing haemophilia (h) is carried in the non-pairing region of the X-chromosome, and is recessive to the normal condition (H). The Y-chromosomes are shaded, the X-chromosomes are not.

from the disease, since the sons will inherit no X-chromosome from their father while every daughter receives H from her mother. However, the daughters are all heterozygotes, capable of transmitting the disease in the manner already described. Here heterozygotes and homozygotes segregate in equal numbers but, once more, relative to sex, since the former type includes males, and the latter females, only.

A marriage between a haemophiliac and a carrier should produce all four types of children in equal numbers: normal sons and haemophiliac sons, normal (but heterozygous) daughters and haemophiliac daughters. It was formerly thought that the condition was lethal in double dose so that there were no records of affected females. But it is now known that such women have survived and even given birth to children; all their sons being haemophiliacs, as expected. The details of this form of segregation will, however, be examined later (pp. 53–5) in relation to a condition more easily studied.

It is highly important to notice that a normal woman can transmit haemophilia while a normal man cannot do so; for he possesses only a single X-chromosome, carrying H. A man who is a member of a family in which haemophilia occurs may marry in confidence that his children will not inherit the disorder if he himself be healthy. In similar circumstances a woman can be given no such assurance. A normal woman who has a haemophiliac brother should realize that the chances are equal that her children would completely escape the disease or that she would pass it on to half of them. Her affected daughters would be healthy though carriers, but her haemophiliac sons would die at an early age. Further, should a maternal uncle be a haemophiliac, she may also be a carrier. Considering the female line only, the chances that she is a heterozygote are evidently reduced by half for each generation since the most recent occurence of the disorder: they are one in four if she has a haemophiliac uncle, one in eight if she has a haemophiliac great-uncle. Yet the existence of normal brothers and uncles, in the female line, in generations *subsequent* to the last occurrence of the disorder may, if sufficiently numerous, fairly dispose of the possibility that a woman is a carrier. It would not be unreasonable to consider her as free from the haemophilia gene if the number of such normal brothers and uncles amount in all to eight or more. Any woman who belongs to a family in which haemophilia is known to have occurred should carefully consider these facts before she contemplates marriage.

The inheritance of haemophilia is illustrated by Queen Victoria and her family. She was herself normal but a carrier of haemophilia. Of her four sons, three were healthy and one (Prince Leopold) was a haemophiliac who actually lived to contract a marriage and produce a daughter and a posthumous son. The occurrence of the disorder among the Queen's descendants has become a matter of history. It will be realized that the present English royal family is not subject to haemophilia since King Edward VII, being normal himself, could not transmit it.

A second and much rarer type of haemophilia also exists in which the symptoms are similar in kind, but milder. Affected individuals are not seriously inconvenienced by their disease and, though slight blows may result in considerable bruising, they are able to lead nearly normal lives. This mild haemophilia is inherited as a sex-linked recessive in the same way as the more frequent severe form. Moreover, since the two types retain their characteristics in the different members of the families in which they occur, they cannot be attributed to an identical gene with variable effects (pp. 79–82). It is reasonable to conclude, therefore, that the mild form is due to another allelomorph at the haemophilia locus.

In von Willebrand's disease, which also affects blood clotting, the capillaries are imperfect and, in addition, though the clotting factor No. VIII, absent in haemophilia, is formed, its concentration is reduced. This situation is inherited as an autosomal 'dominant'. It has been suggested that the manufacture of clotting factor VIII is controlled by this gene, the action of which is suppressed by the mutant at the haemophilia locus. If so, in von Willebrand's disease the suppressor is not operative but the main, autosomal, gene is replaced by an allele which works incorrectly.

Several conditions analogous with haemophilia are known. Thus Christmas disease, in which the symptoms are milder, is due to a deficiency of one of the other clotting factors (No. IX). A mixture of blood from patients suffering respectively from the two defects clots normally. Christmas disease is also a sex-linked recessive, the gene being separated from that for haemophilia by twelve crossover units.

As already mentioned, certain aspects of sex-linkage cannot easily be studied on haemophilia, in which few affected females are known. They can, however, be examined in a far milder and commoner affection: red or green colour-blindness. In England this occurs in about 8 per cent of males and in about 0·64 per cent of females for, with total sex-linkage, the frequency of a recessive character is distributed in men and women as $q:q^2$, attributing equal survival values to the two genotypes (see p. 116). It is inherited in the way just described, consequently only those few features not previously explained will now be considered. In reality, however, the condition is controlled by two loci, affecting red perception and green perception respectively (pp. 124–6).

If a colour-blind man marries a woman heterozygous for colour-blindness, normal and colour-blind individuals will be distributed equally among both male and female offspring, the normal females being all heterozygotes (Fig. 10). When colour-blind females marry normal males, all the sons naturally receive their only X-chromosome from their mother and this carries the gene (c) for

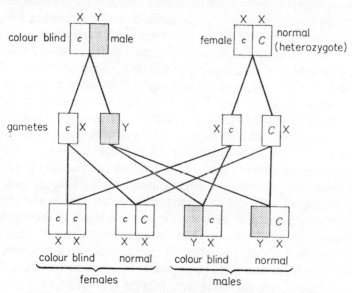

FIG. 10. Total sex-linkage in the X-chromosome: a marriage between a man with a recessive character and a heterozygous woman. The gene producing colour-blindness (c) is carried in the non-pairing region of the X-chromosome, and is recessive to the normal condition (C). The Y-chromosomes are shaded, the X-chromosomes are not.

colour-blindness. The daughters obtain an X-chromosome from both parents, that from the father transmitting the normal allelomorph (*C*), dominant in effect. Thus all the sons are colour-blind like their mother, and all the daughters are normal like their father (though they are heterozygotes): 'criss-cross' inheritance. On the rare occasions when two colour-blind parents marry, all their children must be colour-blind. There is some evidence that the gene for colour-blindness is not fully recessive, so that sometimes the condition is at least partly expressed in the heterozygotes. Its frequency in women is therefore slightly above expectation.

The genetics of colour-blindness emphasize the important fact that there is no inherent barrier to the occurrence of a sex-linked character in both sexes. This is in contrast with sex-controlled inheritance (p. 81), in which the characters concerned are limited to one sex for physiological reasons. However, sex-linked characters appear more frequently in one sex than the other: if recessive, they are commoner in the male while if dominant, in the female. For a gene recessive in effect manifests itself whenever present in the male, but in the female it is dependent for expression upon the slender chance of distribution to the two X-chromosomes of the same individual. With dominant sex-linkage, on the other hand, the presence of the two X-chromosomes in the female allows approximately twice the opportunity for the occurrence of the gene compared with the male. Mutants of this type appear to be exceptional, but that producing G6PD deficiency (pp. 108-9) provides an example of them. Its behaviour can readily be worked out from the information already provided in this section, remembering that the heterozygotes will be included in the classes manifesting the defect. However, it is worth while to draw special attention to two of the situations to be revealed when this analysis is undertaken. First, if an affected man marries a normal woman, criss-cross inheritance in the opposite direction to that described above occurs: all the daughters inherit the defect while all the sons are normal. Secondly, when an affected (heterozygous) woman marries a normal man, normal and abnormal sons, normal and abnormal daughters, will be produced in equal numbers; for all the children receive one or other of the maternal X-chromosomes. Now it is important to notice that this particular result is not distinguishable from dominant autosomal inheritance, though the transmission of the gene in other circumstances will clearly indicate its sex-linked nature.

It will be realized that all the normally sex-linked genes, being those situated in the region of X not homologous with material in Y (Fig. 6, *b–c*), are linked with one another. Crossing-over between them is of course possible in the female, in which this section appears twice, but not in the male, in which it occurs once only. A common affection such as colour-blindness can be used in such linkage studies with the rarer sex-linked characters, and the linkage relations between it and haemophilia have been estimated (Bell and Haldane, 1937). The cross-over value proved to be about 5 per cent, so that the two genes are situated close to one another (and see p. 144 for the map-distances on the X-chromosome).

(*b*) PARTIAL SEX-LINKAGE

Genes situated in the presumed homologous regions of the X- and Y-chromosomes (Fig. 6, *a–b*) differ genetically from the ordinary sex-linked type, discussed in the last section, in two respects. First, they can be transferred between these two chromosomes and, secondly, they possess allelomorphs in both sexes, instead of in the female only. These are of course equal consequences of the existence of homologous material at identical loci in both types of sex-chromosome, but they have somewhat different genetic effects. Were their crossing-over entirely unimpeded, so as to give a C.O.V. of 50, such genes could not directly be distinguished from those carried autosomally. If, on the contrary, crossing-over between X and Y were very rare, so as to escape observation in ordinary investigations, a dominant partially sex-linked gene situated in X could not be distinguished from a dominant showing total sex-linkage.[1] In similar circumstances, a recessive could not be confused in this way owing to the second of the two propositions just stated: that it will possess an allelomorph in Y as well as in X.

[1] When situated in Y it might be confused with total sex-linkage in that chromosome, see p. 57.

55

We may analyse the situation in a recessive partially sex-linked condition. Affected individuals receive the gene responsible for it from both parents. In the mother it must be carried in an X-chromosome, in the father it may be carried either in X or in Y. Consider these two alternatives. If no crossing-over occurred, half the daughters but none of the sons would be of the recessive type if the gene is carried in the paternal X-chromosome; but if in the paternal Y, the reverse would be true. But crossing-over may transfer the gene from one paternal sex-chromosome to the other, leading to exceptions: that is, to the appearance of the recessive character in occasional sons on the one hand or in occasional daughters on the other.

Thus recessive partial sex-linkage will superficially resemble the autosomal system. From this it can be distinguished by an association with sex *within given families*. In some of these, affected males will be rare, and in others affected females. Thus it is only by the analysis of individual family records that such inheritance can be separated from the autosomal form. Were the cases in the population as a whole to be pooled for study, the excess of males and of females would balance each other, so that no indication of sex-linkage could be obtained. Partial sex-linkage can now be illustrated by two examples.

(1) Pseudoxanthoma elasticum is a disease characterized by skin lesions, together with visual and vascular abnormalities. The skin lacks elasticity and may form redundant folds. Choroido-retinitis is present and gives rise to grey bands running across the retina. It is claimed that this feature can sometimes be detected in the heterozygotes. Moreover, haemorrhages occur from the smaller arteries, particularly of the gastro-intestinal tract. It seems that the underlying defect is due to changes in the collagen, and that the heredity is that of a nearly complete recessive showing the characteristics of partial sex-linkage.

(2) Xeroderma pigmentosum is a condition in which the skin is exceptionally sensitive to light, owing probably to some metabolic disturbance. It is usually recognizable a few weeks or months after birth. Exposure to sunlight causes reddening of the face and hands. This is followed by the development of severe freckles

which do not disappear even when the skin is shielded from the light for long periods. These freckles become larger, more numerous and more deeply pigmented. Later, warts arise and patches of skin become atrophic; while corneal ulcers make their appearance, to be followed by opacities. The patients shun strong light, and though the skin may be unhealthy over most of the body it becomes grossly abnormal only in the exposed areas. All attempts to ward off the ultimate consequences of this disease are unavailing: sooner or later multiple malignant changes take place in the skin, the conjunctiva, or the cornea. These usually give rise to basal-celled carcinomas with only a very small tendency to metastasize. They can be kept in check for a time by repeated operations and radium treatment, but eventually they become uncontrollable and the patient dies of cancer of the face.

It appears that xeroderma pigmentosum is not quite recessive, for there is evidence that the heterozygotes may develop exceptional freckling. It is doubtful if a very heavily freckled individual should have children if a member of a family in which xeroderma pigmentosum has occurred: unquestionably such a person should not marry a relative (pp. 190-2).

(c) TOTAL SEX-LINKAGE IN THE Y-CHROMOSOME

The non-pairing region of the human Y-chromosome is of course inherited solely down the male line and there is nothing comparable with that situation in women. In man, Y is the smallest of all the chromosomes so that it must carry relatively few genes; nor can any of those totally restricted to it be essential to normal life since the female must survive without them. Among these in man, but not in the majority of organisms, are those predominantly responsible for male sex-determination (p. 41-2), while another totally Y-linked gene produces hairy ears.

It will be noticed that the existence of sex-linkage affects very little the statement that the two sexes contribute in approximate equality to the heredity of their offspring. The pairing sections of X and Y are autosomal and have no bearing on the matter. There seem to be very few genes totally sex linked to Y. Moreover, the effects of mutant genes in the non-pairing sections of X appear with greater frequency in males if recessive and in females if detectable in the heterozygotes, so tending to reduce any sexual bias introduced by the sex chromosome mechanism into the second proposition of genetics (p. 25).

2.3 THE SEX-RATIO

The simple distribution of the X- and Y-chromosomes would seem to ensure equality in the number of male and female conceptions. This expectation is not fulfilled and, indeed, the relative proportions of the two sexes vary at different times of life and in different circumstances. These facts, when analysed, throw light upon several social and medical problems. In preparing the following survey of them I have drawn largely upon the account of Crew (1937), to which reference should be made for more detailed information.

The sex-ratio is usually expressed either as the number of males to one hundred females, or as the percentage of males in the population concerned. Thus a sex-ratio of 100, or of 50 per cent, indicates numerical equality. Consequently, males are in excess when the sex-ratio is 'high' (over 100, or above 50 per cent), and females are in excess when it is 'low' (under 100, or below 50 per cent). The sex-ratio at conception and at birth is called the 'primary' and the 'secondary' sex-ratio respectively. That at sexual maturity is often referred to as the 'tertiary' sex-ratio in other animals, but in man the expression is too inexact to be useful.

The Report of the Registrar-General for 1935 shows that in England and Wales there is a slight excess of male births, the secondary sex-ratio being 105:100 (51·2 per cent). This is progressively reduced, so that numerical equality is reached in those aged from 15 to 19 years. In the next five-yearly age-group, of 20 to 24 years, females predominate for the first time, and they become proportionately commoner in the population until there are actually twice as many women as men among those aged 85 and over.

The differential elimination of males indicated by these data is apparent also in pre-natal life, for the sex-ratio of foetuses dying during the seventh to ninth months of intra-uterine life is higher than at birth, being 110 (or 52·3 per cent). Indeed there is evidence that this is not only maintained but intensified in passing backwards towards conception. Thus the sex-ratio of infants dying during the first year after birth is highest for the very young while, according to Crew (1937), it is higher also among early than among late abortions. We may fairly conclude, therefore, that the primary sex-ratio departs widely from equality in the direction of an excess of males.

It will now be convenient to study somewhat further the results of this trend, then to seek an explanation for it, and finally to consider the problem presented by the high primary sex-ratio.

The lower viability of males than of females is responsible for the way in which the sex-ratio differs in varying circumstances since, within certain limits, unfavourable conditions tend to accentuate and favourable ones to mask it. There is ample opportunity for the secondary sex-ratio to be affected in this way since, on the average, only 76 per cent of conceptions yield living offspring.

The secondary sex-ratio is highest among first-born children, since abortion and miscarriage are relatively more frequent in large than in small families. It is lower among the children of old than of young mothers, for with advancing age the internal environment becomes less favourable for the foetus. As abortion is more frequent in urban than in rural communities, the secondary sex-ratio is lower in towns than in the country. For obvious reasons also, abortions and still-births are particularly common in respect of illegitimate children, whose secondary sex-ratio is consequently low. It is often stated that a relative increase in male births takes place in time of war. If this be a fact, and the evidence for it does not appear to be conclusive, it is possibly due to the wider spacing of births owing to absentee husbands. For repeated conceptions, following too rapidly upon one another, create an unfavourable environment for the developing foetus. However, any detectable effect of this kind appears rather to follow the conclusion of hostilities. Thus it may well result from the fact that there is an excess of first-born children at this time (who, as already mentioned, have a high sex-ratio), owing to the frequent postponement of marriage or of child-bearing during war. All factors which tend to reduce the standard of living tend also to lower the secondary sex-ratio, which is higher in the upper and middle classes than among unskilled labourers. Consequently, as stressed by Crew, changes in the secondary sex-ratio can be used as an index of the effectiveness of social services, such as slum clearance and attempts to improve the general health of the community.

All these facts are dependent upon the higher death-rate of males than of females, the basis of which merits inquiry. Clearly, the more exposed and dangerous lives led by men than by women must contribute to the excess of the latter in the higher age-groups, though it must be remembered that deaths at childbirth tend heavily in the opposite direction. But such considerations are inapplicable to infancy and to pre-natal existence, so that other and more fundamental causes must be sought for the phenomenon.

The XX, XY chromosome mechanism itself constitutes a system which acts unfavourably upon the heterogametic sex. As will be explained in Chapter 4, disadvantageous genes tend to become recessive in effect. Now if we exclude the small section of X homologous with Y, it can be said that recessive sex-linked characters are always expressed in men but very rarely in women. The reason for this is obvious. The recessive sex-linked genes in a single X-chromosome cannot be opposed by a dominant partner, as they will usually be when two X-chromosomes are present (see p. 53). The region of X which is not homologous with Y must contain many hundreds of genes, a few of which will almost certainly be in the recessive state in any individual and capable of lowering his resistance to disease or reducing his efficiency in other ways.

It will be noticed, however, that these considerations apply not to the male as such but to the heterogametic sex. Yet it is found that the male is still the more heavily eliminated in those forms (e.g. birds and Lepidoptera) in which the female is the XY type, but to a less extent than in the reverse condition (Crew, 1937). This suggests that while sex-linkage is a cause of the lower viability of the male in man, it is probably not the only one. Some additional agency, a function of sex itself, must also contribute to it, and very possibly this is to be found in the higher basal metabolism of men than of women, which affects both the young (Benedict and Talbot, 1921) and the adult (Benedict and Emmes, 1915). It is probable that the greater expenditure of energy by the male makes him less resistant and more liable to death than the female, a view which is supported by various experiments on lower forms (see Crew, 1937).

It is clearly necessary to explain the heavy excess of males at conception, despite the operation of a mechanism which appears so certainly to ensure the numerical equality of the sexes at that time. In order to understand this, it must be remembered that genes are well known which act differentially upon the X- and Y-bearing gametes produced by the heterogametic sex. One of them studied by Gershenson (1928) in the fly *Drosophila pseudo obscura* may specifically be mentioned as it provides an extreme instance of the kind. It eliminates nearly all Y-bearing sperms and is quite common in nature. It is clear that many genes affect unequally the survival of the two types of sperms so that the sex-ratio at fertilization is susceptible to selection. In man this operates to produce approximate equality of the sexes during the period of life when reproduction is chiefly taking place: about 16 to 30 years of age. In view of the substantially higher survival of the female sex before and after birth to which attention has been drawn, such equality can only be obtained by favouring a heavy excess of male conceptions.

Allan (1972) has produced evidence to show that the sex-ratio at birth is influenced by the ABO blood groups (*q.v.*). He finds that this is significantly higher for group AB mothers than for those of groups O, A and B combined; that the sex-ratio of B babies of B mothers and of O babies of O mothers is higher than that of A babies of A mothers, and that it is higher also for O babies as a whole than for A babies as a whole. On the other hand, it is lower for non-B babies of B mothers, and for non-O babies of O mothers than for non-A babies of A mothers. Thus there is here a distinction between mothers whose babies are of the same ABO group as themselves and those whose babies are of another ABO group.

CHAPTER 3

Mutation

3.1 INTRODUCTION

Mutation may be defined as the inception of a heritable variation. Changes in the genes themselves, or in the chromosomes which carry them, may be reflected in the characters of an organism, causing it to differ from the normal form. Since such changes affect the hereditary material they can be transmitted to subsequent generations, so that the conditions which they produce are inherited. It will be evident that mutations may not have an immediate influence, as when one member of a homologous pair of allelomorphs mutates to produce a gene recessive in effect. Many generations may then elapse before a homozygous recessive arises, so that the appearance of an individual possessing a novel recessive character may indicate only the occurrence of a mutation in the distant past. On the other hand, an individual possessing a dominant character new to the stock which is being investigated usually demonstrates that a mutation has taken place in the germ-cells of one of its parents.

It is desirable to adhere strictly to the definition of mutation adopted here. Mutations are sometimes defined as changes in the genes; yet this is far too restricted, for chromosomal modifications lead also to inherited variations. Furthermore, the concept that some diversity exists in the germinal material is implied in all theories of inheritance, and so must be the inception of such diversity. It should therefore be possible to discuss the part assignable to mutation in the various hereditary systems postulated at one time or another, and to compare it with that which it is now actually known to play.

It is unfortunate that the term mutation is often used in two senses: first, for the actual change in the hereditary material;

secondly, and quite incorrectly, for the result of that change. For instance, the less precise writers on genetics quite commonly refer to a gene which has arisen during the course of experimental work as a mutation. For this, the expression 'mutant gene', which may be said to give rise to 'mutant characters', may be employed, and is unexceptionable. 'Mutation' and 'mutant' may indeed be compared with the terms 'conception' (of an idea) and 'concept'. Further, the word mutation is one which long antedates the development of modern genetics and, like many others now accurately standardized, has carried with it a tradition of inexact usage. Thus a novel *character* was originally described as 'a mutation', though it might in reality be due to recombination or even to environmental variation. However, when a term has acquired a definite technical meaning, the fact that it was once employed less strictly is not an excuse for using it loosely today. Were this so, current nomenclature both in biology and in medicine would have to undergo a far-reaching and wholly unnecessary revision. It should now be clearly appreciated that neither a gene nor a character can be called a mutation; only the act of change in a hereditary unit can so be described.

It has already been indicated that mutations affecting the hereditary material as we know it can be subdivided into changes in the genes and in the chromosomes. These two aspects of the process must briefly be considered.

3.2 CHROMOSOME MUTATION

Chromosome mutations can be grouped into two main classes: producing abnormalities in the distribution of whole chromosomes, and chromosome fragmentation. The first of these can again be sub-divided into *polyploidy*, in which the entire set of chromosomes is multiplied, and *heteroploidy*, in which a chromosome may be added to or lost from a single chromosome-pair.

Polyploids arise from failures in cell-division during meiosis, leading to the formation of a gamete with the diploid (unreduced) chromosome number. This, on fusing with a normal haploid gamete, produces a 'triploid' zygote possessing three, instead of two, sets of chromosomes. Further irregularities, such as the fusion of two diploid gametes, may lead to the production of individuals with four chromosome sets (tetraploids) or even higher chromosome values. As Darlington (1956) points out, polyploidy almost always increases, hardly ever decreases, chromosome numbers. For though haploid individuals may sometimes arise possessing a single chromosome set only, as in the gametes, this occurs by a different means, through 'parthenogenesis'. A spermatozoon which fails to achieve fertilization will occasionally stimulate an egg to develop, and the individual which it produces will possess a single maternal chromosome set only. Polyploids have not been encountered in man and indeed they are always rare in animals, though they are important and widespread in plants. For a new polyploid, being sterile with the parental form, will generally need vegetative reproduction for the early stages of its spread.

Heteroploids cannot very often establish themselves, for the addition or loss of a single chromosome may have a greater and more harmful effect than the addition of a complete set. This is due to the lack of genetic balance which ensues from quantitative variations in part only of the hereditary material. Heteroploidy of the sex-chromosomes, giving rise among other conditions to the Klinefelter and Turner syndromes, has already been mentioned, but it has been detected in the human autosomes also. Thus Mongolian idiots possess an extra member of one of the chromosomes, giving a count of 47, these being the twenty-first when the set is numbered downwards according to size. That condition produces Mongolism in certain environments only, for it does so increasingly as maternal age advances and produces a less favourable situation for the foetus. In Western Europe the frequency of this type of mental deficiency is one in 2,000 births for the maternal ages of 29 and under; when the mother is 45 and over, it is one in 54. Moreover, trisomy in chromosome number 17 as well as within the group numbered 13-15, produces mental retardation associated with physical features of another, and much rarer, kind, differing from each other and from Mongolism. There is good evidence that trisomics of chromosome 21 are unduly liable to malignancy (Harnden et al., 1971).

Chromosome fragmentation may lead to the loss of part of a chromosome (deletion), the portion which remains is that which contains the spindle attachment. However, the detached piece may re-attach itself, the wrong way round (inversion), or else to the homologous chromosome (duplication) or to a non-homologous chromosome (translocation). Carter *et al.* (1960) indeed demonstrated that there is a second and much rarer form of Mongolism due to a translocation of a fragment of a small chromosome (number 21) to a larger (number 15).

As well as translocation, an interchange of fragments may occur between non-homologous chromosomes (structural interchange). This may have little physiological effect since the number of genes carried by the individual is not altered. It does, however, produce linkage between those heretofore assorting independently (p. 29).

It is important to notice that in other animals, and in plants, numerous combined cytological and genetic studies of the results of chromosome mutations have been made. For instance, when genes have been found associated with the wrong linkage-group it has often been possible to see that one chromosome is too short and another correspondingly too long, owing to the translocation responsible for the observed genetic irregularity. Such observations provide a remarkably complete demonstration of the chromosome basis of heredity (see Sinnott, Dunn and Dobzhansky, 1958). However, Stern (1931) provided the ultimate proof that crossing-over is due to a sectional interchange of material between two of the chromatids from different but homologous chromosomes in the four-strand stage. He did so by studying two sex-linked genes, carnation and Bobbed, in a strain of *Drosophila melanogaster* in which both X-chromosomes were visibly distinct.

Very short deletions may give much the effect of mutant genes, nor can they be detected cytologically. On general grounds it is not to be expected that the majority of genic changes are due to an actual loss of material, and there is precise evidence for this view.

3.3 GENE MUTATION

Turning to gene mutations, it should be noticed that these are strictly localized phenomena. The occurrence of a mutation at one locus does not affect the allelomorph, nor does it tend to produce mutations at the loci on either side of it. This latter statement may possibly have to be qualified, since a modification in the effect produced by the genes near the locus of a mutation has been reported in *Drosophila* and a plant, *Oenothera*. On the other hand, it does not yet appear to have been established that such genes have also mutated. The alterations in the characters which they produce may be due to a localized interaction between them and the new mutant, 'position effects': indeed it seems likely that this is so.

There is evidence that mutations are slightly more frequent during cell-division than in the resting-stage, but they are not in fact restricted to any one period of cell activity. Nor are they confined to the germ-tract, but can occur in the body-cells also (*somatic-mutation*). This is more important in plants, where somatic tissue can give rise to germ-cells throughout life, than in animals; but it has been widely established in both kingdoms. In man, the process is of potential importance since it may possibly initiate a malignant change (p. 97). An example of it is probably provided by heterochromia iridis.

Of great theoretical significance is the fact that *reverse mutation* can occur, from the mutant gene to the allelomorph from which it was derived: from the abnormal to the normal. It is clear, therefore, that mutations are not essentially destructive changes, from more to less highly organized states, or downwards from compounds possessing high chemical energy. This fact assures us that mutations as we encounter them are of a type suitable to initiate the genetic diversity of living organisms.

Though mutations are exceedingly rare phenomena, yet they are recurrent, and take place with characteristic frequencies, being relatively commoner at some loci than at others. It is probably true to say that one mutation in a million individuals is an approximate average mutation-rate for a given gene. A mutation-rate of

1 in 50,000 individuals is exceptionally high and rarely exceeded.

These values appear to be approximately similar throughout the widest range of organisms, both animals and plants, and for this there is good reason. We may be confident therefore that they apply also to man. It is, however, a somewhat remarkable fact that we actually possess direct evidence of the human mutation-rate in respect of a few genes. The lines upon which such information has been obtained must briefly be indicated.

The first estimate of mutation-rate in man was obtained by Haldane (1935) from a study of haemophilia. It has been pointed out that this is a recessive sex-linked disease of so severe a nature that nearly all haemophiliacs die without producing offspring (pp. 49–52). Two thirds of the X-chromosomes of the human race occur in women and one third in men. One third of the haemophilia genes in the population are, therefore, always exposed to selection, and are nearly always eliminated, while the remainder are not. Thus were they not supplied anew by mutation, the disease would rapidly die out, and at a calculable rate. The frequency of mutation necessary to maintain haemophilia in the population can therefore be found. Haldane was able to show that the proportion of haemophiliacs living in Greater London in 1935 must lie between 35 and 175 per million male births. We may be confident that this proportion is in equilibrium, a conclusion easily accepted if we consider, with Haldane, the situation in former times were this not so. Suppose mutations were not making up for the known elimination of haemophiliacs, so that the haemophilia genes now existing were derived from haemophiliacs or heterozygotes in an earlier population. Thirty generations ago haemophilia would be 100,000 times commoner than it is today, and the whole male population of England would be haemophiliacs at the time of the Norman Conquest. Haldane calculates that one mutation in 50,000 individuals is required to balance the elimination of the haemophilia gene, which therefore represents its mutation-rate.

The study of epiloia has provided material for an investigation of mutation at another locus (Gunther and Penrose, 1935). This disease is one in which mental deficiency is associated with adenoma sebaceum and other skin defects (pp. 77, 78). It is inherited as an autosomal dominant, yet a number of cases are known to have had normal parents. It is highly unlikely that these represent failures in heterozygous manifestation, since the segregation is always very clearly marked. Such sporadic occurrences must, in fact, be mutations, and Gunther and Penrose find that they take place with a frequency which lies between one individual in 60,000 and one in 120,000 of the population.

It is most important to notice that such estimates of human gene-mutation do not represent an average but approach the upper limit of its frequency. For evidence along these lines could not be obtained for mutation-rates much lower than those observed. It is evident, therefore, that gene-mutation in man is of the same order of magnitude as that in other organisms. In those instances actually studied, the frequency is slightly higher than of the most mutable genes of *Drosophila* (the gene for white-eye, with a mutation-rate of one in 80,000 individuals), but does not fall very materially beyond this. Nevertheless, some indication of the more normal mutation-rates in man can occasionally be obtained. The males affected by mild haemophilia, which is allelomorphic to the more severe form, are handicapped to a comparatively slight degree and often become parents, though they must contribute somewhat less to posterity than do healthy individuals. Thus the selection operating against this gene is far less severe than that against ordinary haemophilia, yet the mild type is much the rarer. Consequently mutation to this gene must be very much more uncommon than to the other haemophilia gene, and a frequency of one in a million individuals is possibly a fair estimate of it.

Mutation must normally be due to the fact that, in general terms, the copying process involved when the genes multiply in the resting stages prior to mitoses and the prophase of the first meiosis cannot invariably be exact. Its frequency increases with temperature but to an extent which it is difficult to quantify. However, its occurrence seems to be multiplied by a factor of 3 to 5 for a rise of 10°C.

H. J. Muller was the first to show that mutations can be induced by X-rays. In favourable circumstances, the mutation-rate can be increased about 150 times by this means. It is now established that all types of shortwave radiation are capable of producing mutations: X-rays of different lengths, the γ-rays of radio-active substances, free electrons (β-rays of radium), and neutrons. These agents are all equally effective so long as equivalent doses (judged by ionization rates) are employed. Special allowance must of course be made for those of low penetrating power. All forms of mutation, both genic and chromosomal, can be induced by these means, but entirely at random. As with the spontaneous type, they do not cause allelomorphic genes, or those at neighbouring loci, to mutate. No one agent tends to produce a given class of mutations, or to induce them at one locus rather than another, nor can the results be controlled by varying the dosage or by any other known means. Many of the mutations which take place are the same as those which have appeared spontaneously. Others have not been observed except as the result of artificial induction, but it is highly improbable that these are, in reality, novelties. As such treatment increases the frequency of mutations so greatly, some will naturally be encountered which have not yet been detected in normal circumstances. Furthermore, reverse mutation can be induced by short-wave radiation.

We are concerned here solely with the genetic action of penetrating radiation; that is to say, its mutagenic properties: no reference will be made to its important and profound clinical effects. Mutation of all kinds is very rare because selection has had to ensure that highly stable molecules are employed for the genetic material, which also is, as far as possible, protected from environmental effects.

Putting it another way, much energy is required to produce mutation. This is particularly true of chromosome breakage which, relevant to ionization, can only be materially increased by heavy dosages. Even so, it will not often be available for study as individuals receiving a broken chromosome are very likely to die. It will be appreciated that we are not here concerned with errors in cell-division, leading to polyploidy. These can be produced experimentally by injury and the use of drugs.

Point mutations are normally due to imperfections in the copying process when the genes reproduce. It is not clear what agencies are responsible for such occurrences, but background radiation is thought to produce 2–20 per cent of them. About 28 per cent of this is due to cosmic rays, derived from the stars, and the remainder to radioactive materials in our environment: potassium[40] and traces of the heavy elements.

Mutations will arise when radiation removes an electron from an atom in the hereditary material. Their frequency is directly proportional to the dosage received, judged by the ionization rate. It will be appreciated, however, that the amount delivered to the germ-cells depends upon the penetrating power of the rays to which the organism is exposed: very great with γ-rays, very slight for α particles; though even the latter can be important when they arise from radioactive substances absorbed into the body. Evidently in mammals the ovaries are better shielded by the tissues than are the testes. Ultra-violet light can also produce gene, but not chromosome, mutations; while its action is not directly proportional to dosage.

It has long been known that polyploid shoots can be produced in plants by the application of certain substances, in particular colchicine. This has the advantage of being water-soluble and is non-toxic at the concentrations required. It acts by a direct effect upon cell-division. It has been used successfully also to produce polyploidy in animals (rabbits). However, it was not until 1947 that Auerbach and Robson discovered a powerful chemical mutagen. This is mustard gas $2(Cl.CH_2.CH_2)S$, which is almost as effective as the highest doses of X-rays that can be used without sterilizing *Drosophila*. It has subsequently been found that certain other substances (for instance, nitrogen mustard) are also strongly mutagenic.

Mutation is predominantly harmful since it is randomized relative to the needs of the organism, and random changes in a highly adjusted system can seldom promote its harmonious working. The chances are a little increased that a new mutant may give some slight advantage in a new environment, genetic or external, as it has been found to do in the domestication of animals and in the cultivation of plants. Nevertheless organisms must in general protect themselves from mutation and should therefore be protected as far as possible from exposure to ionizing radiation. Moreover, from what has been said it will be noticed that there is a distinction here compared with the clinical situation. For long exposure to small amounts of penetrating rays being cumulative in effect can do as much genetic damage as a single dose so heavy that it could never be permitted to hospital nurses, radiologists or workers in atomic energy stations.

It has now been possible to induce mutation by means of short-wave radiation in a great variety of organisms both plants and animals, including insects (*Drosophila, Habrobracon* and others) and mammals (mice). They all respond in a similar way. Indeed the uniformity of the effects produced by penetrating radiation on the widest variety of organisms allows us to generalize with confidence on their influence on the human hereditary material.

Satisfactory data do not yet appear to be available on the proportion of miscarriages, abortions and still-births among the offsprings of radiologists. Despite the protection afforded by modern technique, it is hardly possible that these conditions are not in some slight excess when compared with the rest of the population. They would, of course, arise from the production of mutant genes dominant in effect. A very differing situation is presented by the recessives resulting from radiation mutations, and this matter has been studied by Muller (1941). The appearance of the homozygous recessive type may be due either to the mutant gene meeting another like it, which had been induced independently, or to the meeting of two genes descended from the same original mutant. Muller investigated all the aspects of these two possibilities, including the frequency of induced mutation, the survival-value of abnormal types, and the degree of inbreeding in human populations. He reached the conclusion that even in the most closely inbred groups of modern society, a gene recessive in its effects originating from radiation technique would remain latent for 600, and probably for several thousand years. Some potential danger from the accumulation of such genes in the population none the less exists, but it cannot of course outweigh the diagnostic and therapeutic benefits of irradiation. Yet it is well to draw attention to a fallacy which frequently finds support when such a situation as this is exposed: that the accumulation of harmful recessives arising by mutation, due in this instance to radiation therapy, can have a deleterious influence upon human evolution. Under Mendelian inheritance, evolution is not controlled by mutation.[1] The use of radiation today may increase the proportion of disadvantageous types segregating in the distant future. These will be subject to counter-selection as they appear, but they cannot influence the course of evolution. However, when we consider the possible consequences, both immediate and remote, of the fact that short-wave radiation induces mutations, the importance of shielding the gonads during radiation intended for other parts needs no additional emphasis.

It has been pointed out that there are good grounds for supposing the average human mutation-rate to be approximately the same as in other animals and in plants, and that even at its highest it can hardly exceed that of the more mutable genes in the fruit-fly *Drosophila* (p. 68). This uniformity of the mutation-rate in organisms having greatly different lengths of life is a highly remarkable and significant fact. As already mentioned, mutation is not restricted to any one period in cell activity. Consequently it should be measured per unit of time; its frequency per generation should be irrelevant. Yet this perfectly logical expectation is not fulfilled. Haldane (1935) showed that if we take fourteen days and thirty years as reasonable average lengths of a generation in *Drosophila* and man respectively, the upper level of human mutation of about one in 50,000 (for the haemophilia gene) when adjusted for time is equivalent to a mutation-rate of one in 40 million per *Drosophila* generation: a value so minute as to fall outside anything we know of mutation in that organism. Judged per generation, man has nearly the same mutation-rate as *Drosophila*; judged per year, *Drosophila* is immensely more mutable than is man.

We may well marvel at the astonishing relation between the mutation-rate and the generation, irrespective of its length, revealed by these facts, yet its explanation is not difficult to find when we recall that all inheritance is Mendelian. The great value

[1] See Fisher (1930*a*).

of this system, when compared with any blending mechanism, is the permanence of its hereditary units. This enables selection to favour even minute advantages and allows an organism to make use of genes which arose far apart in time or in space, so escaping the 'swamping effect of intercrossing' which proved so great a stumbling-block to Darwin, on his assumption of blending (Fisher, 1930*a*). But this crowning advantage of the particulate system is wholly discarded if mutation be common. The vital permanence of the genes is then lost, even though they do not contaminate each other. With Mendelian inheritance, mutation not only need not, but must not, be frequent. Selection must therefore restrict the occurrence of mutation *per generation* sufficiently to ensure a high degree of permanence to the genes. Its average admissible limit will be a given frequency per generation, to ensure which the frequency per time between organisms having long and short life-cycles must be vastly different. Thus the human mutation-rate is in accord with the Mendelian, and with no other, system of heredity: while its similarity per generation to that of other organisms demands the fundamental action of selection in a manner which must affect profoundly, and in diverse ways, the analysis of human heredity.

The permanence of the genes would be lost were they to contaminate one another. It would be lost also if they had a high mutation-rate (pp. 27–8). Thus the *control* of evolution by mutation is impossible; and so are those theories which involve that concept, such as 'Lamarckism' and the inheritance both of 'Acquired Characters' and, in general, of the effects of the environment upon the body. Those views, long untenable, are the more absurd now that the power of selection proves to be immensely greater than was supposed twenty-five years ago (Ford, 1971). The suggestion that protein variability is maintained at a high level by mutation, producing genes of neutral survival value, is discussed on pp. 114–16.

CHAPTER 4

The Action of the Genes

The previous chapters of this book have been restricted to a discussion of the behaviour of the genes, their segregation and mutation, and of the effects to which they give rise. For this purpose particular genes have been treated merely as the inherited basis of given characters. It is, however, important to bridge the gap between genes and characters, and to determine what type of effects the genes may have and the ways in which they produce them: to obtain, in fact, some information of the physiology of genetics.

4.1 THE EFFECTS OF THE GENES

Certain types of genetic effect may usefully be classified before attempting to consider the behaviour of the genes in varied environments. It will then be possible to give a brief account of the dominance phenomena, and of some instances in which genetic action has been analysed from the physiological point of view.

Attention will repeatedly be directed to the fact that distinct genes may have nearly, or completely, identical effects. Thus retinitis pigmentosa may result from the operation of five, or more, different genes. Of these at least two are autosomal, one being dominant and the other recessive in expression, one is recessive and totally sex linked, while two, one dominant and one recessive, are probably allelomorphs. Only the autosomal recessive is distinguishable in its effects since, unlike the others, it produces deafness in addition to the characteristic changes in the eye. A number of similar instances might be given; such as keratosis follicularis (Darier's Disease), which has been produced by a sex-linked gene, though the disorder is usually inherited as an auto-

74

somal dominant. However, it appears unprofitable to reduplicate examples save in special circumstances. Distinct genes may also produce effects which are very similar in kind but different in degree, so resembling the series of characters to which multiple allelomorphs give rise (p. 23). Thus three forms of epidermolysis bullosa are characterized as mild, dystrophic and severe dystrophic. Yet these are not the result of allelomorphs but are due to three genes at least one of which is known to occupy a distinct locus. The first two are autosomal dominants, that producing the dystrophic form being rather irregular in expression, while the third, responsible for the severe dystrophic condition, is recessive.

The instances so far given of genes having similar effects relate to rare abnormalities, and I know of no occasions on which they have been brought together in the same individual. Were this to occur, several distinct possibilities arise, but only two of them require consideration at this stage: that is to say, the results might or might not be cumulative. Numerous instances are known in animals and plants in which characters due to the operation of distinct genes have been combined, sometimes without increasing their effect. Thus the 'rex' rabbit is one in which the stiff 'guard-hairs' are greatly reduced, producing a plush-like coat of commercial value. The character is recessive and it may be due to the action of any one of three genes, two being linked and one carried in a different chromosome. Their effect is the same whether one, two or three of the homozygous recessives are present. I am not acquainted with any definite examples of the same type in man, though doubtless such exist.

More often, genes producing similar characters are additive in their action. This is undoubtedly very frequent in man, particularly in respect of quantitative differences. An example is provided by normal human height. It can be demonstrated by means of correlation that this is partly hereditary. But marriages between tall and short parents give rise to no clear-cut segregation, owing to the fact that height is controlled by many genes producing small cumulative effects. When a character determined in this way is studied in those animals or plants in which a large progeny can be reared, it is found that the majority of the offspring approximate to the average height while the more extreme forms become progressively rarer, the proportions bearing a definite relation to one another such that the whole group falls within a curve of normal distribution.[1] This applies not only to the F1 but also to the F2 generation, in which the curve remains 'uni-modal'; that is to say, it gives no indication of segregation into different classes. Such a condition is known as 'continuous variation'. It must not be supposed, however, that the inheritance involved is non-Mendelian. Indeed the analysis of heredity undertaken in the latter part of Chapter 1 can admit of no such view. That it is determined by the particulate mechanism is demonstrated by the fact that in these circumstances the F2 generation is more variable than the F1 (pp. 26–7): the inheritance being 'multifactorial'.

It is plain that we have here the occurrence of gene interaction, in the sense that a number of genes, operating quantitatively, are required to produce a given effect. The influence of one gene therefore depends upon the presence of others. But such interaction may take different forms, a well-known variant of it being provided by the *modifying factors*. These are genes which exert no known influence by themselves, so that their existence can be detected only in the presence of others whose effects they modify. Such modifying factors have been studied in great detail in other species of animals and in plants, and doubtless they are as important in man as elsewhere. Indeed general indications of their activity are common in human genetics. It is quite frequently found that a given gene produces somewhat distinct results in different individuals. Albinism, for example, is inherited as a simple recessive, but the cases may depart from the extreme, and typical, form with white hair and pink eyes. Some have pale blue eyes and pale yellowish hair. If such subdivisions of a type appear sporadically, the variations may be environmental (pp. 90–3); if, on the other hand, they are characteristic of particular families, as often in albinism, they are attributable to modifying factors. These may be

[1] This is a bell-shaped curve such that if the ordinate is y, the abscissa is x, and the number of individuals at any point along x be the frequency (f), then log f at any distance from the centre is less than the logarithm of the frequency at the centre by a quantity proportional to x^2.

detected only by their effect on the main gene. In albinism it is probable that some at least of such genes owe their modifying action to the fact that pigment production tends towards colour-saturation, so that pale individuals are more sensitive than dark to a small increase to the total amount of pigment. Often, however, no such simple interpretation is possible, and one gene may act only in the situation provided by the activity of another. Thus the symptoms of epiloia are very variable, and Gunther and Penrose (1935) showed that this is in part due to the operation of modifying factors which affect the expression of the single gene responsible for the disorder. This is usually described as a 'dominant', though it appears to be known only from its heterozygous effect. Similarly, Popenoe and Brousseau (1932) obtained evidence that modifying factors affect the manifestation of Friedreich's Ataxia, though this condition is unifactorial and recessive.

A logical extension of the concept of gene-interaction embraces an exceedingly important group of phenomena constantly encountered in the genetics of animals and plants: that is to say, two genes, each responsible for definite characters, may combine to produce new effects of a distinct type, not merely the average or the sum of those to which they give rise independently. It is for this reason that species-hybrids are frequently found to possess characters absent from both of the parental forms. We may take as an example the fruit-fly, *Drosophila melanogaster*, which normally possesses red eyes. Brown and scarlet eye-colour are both recessive and due to the operation of entirely distinct genes. Yet when these are brought together in the homozygous state in the same animal, they interact to produce white eyes similar in appearance to those due to a sex-linked recessive. The production of distinct effects through the interaction of different genes has so far rarely been detected in man (but see p. 121). However, the most universal nature of the type of phenomenon here discussed, which has been demonstrated in the most diverse forms, leaves no room for doubt that a further search for it would prove successful.

Attention must now be drawn to a fundamental property of genetic action: that single genes are responsible for multiple effects. Several examples of this have already been given incidentally, but the subject is one which merits special consideration. Geneticists, whether they study human material or not, are usually concerned in following the segregation of particular genes. For this purpose it is convenient to select some obvious feature which serves to mark the presence of each: a procedure which tends to obscure the multiple nature of genic action. So much so, that the mistake is sometimes made of regarding each gene as the heritable precursor of one particular character. No view could be more false. We will at present restrict ourselves to discussing that aspect of it which so wrongly associates unit genes with single characters.

Quite possibly, careful examination would reveal that all genes produce multiple effects, often of surprisingly diverse kinds. Even with our present limited knowledge of human genetics, the instances in which these have been detected in man are very numerous, and a few only will suffice as examples. A single gene, described as a 'dominant', is responsible both for bone-fragility and the production of blue sclerotics; while a single recessive gives rise to the Laurence–Moon–Biedl syndrome, the features of which are polydactyly, retinal degeneration, mental deficiency, obesity and hypogenitalism. It has already been mentioned that the form of retinitis pigmentosa due to an autosomal recessive is associated with deafness, in addition to the migration of pigment into the retina (p. 74).

It is an important fact that genes may influence the general constitution to a degree quite out of proportion to their visible effects, and in a manner wholly unrelated to them. For example, albinos are, on the average, shorter in stature and less hardy than are normally pigmented persons. The presence of the gene responsible for epiloia is most easily recognized by the characteristic distribution of the adenoma sebaceum on the face, especially along the naso-labial folds, to which it gives rise. But this symptom bears no obvious relation to the mental deficiency and production of nodules of neuroglial overgrowth in the brain, which make the prognosis of this condition so gloomy. Similarly, nearly all of the genes which have arisen by mutation during the course of genetic experiments are disadvantageous in their effects, though the observable characters to which they give rise are often quite trivial. In *Drosophila*, for example, such genes usually reduce the length of life, and the number of eggs laid per day by the female; yet we may recognize them by the absence of a bristle on the thorax or by a barely detectable change in eye-colour. Nevertheless, their effect on viability makes it clear that they produce other and more profound changes of a physiological kind. The almost universal nature of such alterations in the general vitality of the organism assures us that the action of single genes must in general be multiple.

The facts so far described lead to a conclusion of much conquence. Genes give rise to multiple effects, and they interact with one another to produce the characters for which they are responsible. Therefore, they must combine to form a balanced system, or *gene-complex* (a term introduced by Ford, 1931), which constitutes a form of internal environment within which each gene must act. That is to say, it is not possible to alter any single gene (by recombination or mutation) without influencing the operation of many others. Similarly, the effects of a given gene will be changed when it is placed in a gene-complex whose composition differs considerably from that to which it is accustomed. This situation is, of course, attained when a marriage occurs between those human races that rarely inter-bred, so that a different genetic situation has been built up within each. The effect will evidently be greater the wider the crosses that we study: those between races, species and genera producing progressively noteworthy results. The degree to which we can examine the activity of particular genes in different gene-complexes is, however, strictly limited. Offspring can only arise from matings between fairly closely related genera, and these must have the great bulk of their genes in common. Further, in wide out-crosses the hybrids are usually sterile, and in them each gene is still immersed in a gene-complex to half of which it has been adjusted. None the less, the effects to which a given gene may give rise when placed in a new setting may be dissimilar enough from those which we normally attribute to it. We may take a significant example from the genetics of fish. In the Mexican Topminnow (*Platypoecilus*) a single sex-linked gene *Sp* merely produces dominant black spotting. The hybrids between this and

79

the sword-tail, a member of another genus (*Xiphophorus*), are quite healthy unless they receive the gene *Sp* from their *Platypoecilus* parent, which in the hybrid gene-complex gives rise to a fatal cancerous growth (Kosswig, 1929*a* and *b*).

The forms of mankind surviving today appear to belong to a single species only, nor so far as we know do the various human races differ sufficiently to produce any detectable degree of sterility or sexual abnormality on inter-crossing: and these are the first recognizable signs of those differences which lead to the evolution of races into distinct species. It should, however, be remarked that we have no data on marriages between many of the more dissimilar types of mankind. It is quite possible that some tendency towards heterogametic deficiency or abnormality might be found in matings involving some of them: for example, between Eskimo and Bushman or between either of these and Australian Aborigines. As a reaction against unjustifiable social and political discrimination, there has in some quarters been a tendency to minimize the real distinctions which exist between the white, black and yellow races of mankind (pp. 193–7). To maintain that these differ from one to another 'only in the colour of their skin', a phrase employed by a Judge in an English Court of Law in the last few years, is so obviously untrue that it could not be used by anyone with the smallest knowledge of physical anthropology or, indeed, of medicine (Le Gros Clark, 1965; Dobzhansky, 1966).

However, in the instances so far available for study the amount of modification of the gene-complex which can arise from any racial crossing in man is small compared with that which can be witnessed in wide out-crossings in other animals and plants. Yet we may be confident that further study of human genetics will show clearly enough the influence upon the effect of particular genes which results from their action in somewhat distinct gene-complexes. Nor are instances of the kind unknown today. The woolly hair of the Negro behaves as a 'dominant' to the non-woolly type, but its heterozygous expression is subject to considerable variation in crosses with Europeans. Otto Mohr (1932) analysed an instance in which this gene (or, less probably, one producing similar results) arose by mutation in a pure Norwegian

THE ACTION OF THE GENES

stock. Here the heterozygotes were very constant. The difference in modifiability in the effect of the gene may very reasonably be attributed to the greater variability of the gene-complex among the offspring of inter-racial crosses than can arise from marriages within the same race.

Numerous instances exist in the records of experimental genetics in which a gene, after giving new effects in a gene-complex to which it is unaccustomed, has been restored to that from which it came (Ford, 1940b). In these circumstances, it has produced once more the characters for which it is normally responsible. That is to say, such an alteration in effect is due not to a change in the gene itself but to a change in the response of the organism to that gene, consequent upon recombinations in the set of genetic factors with which it has to co-operate.

A well-known instance of the fact that the genes produce different effects in different internal environments is provided by *sex-controlled inheritance*.[1] This is a condition in which the action of a gene is only, or more frequently, detectable in the environment provided by one of the sexes. Thus it is carefully to be distinguished from sex-linked inheritance (pp. 47–57) in which the relation between genetics and sex is of a mechanical kind; some genes being carried in the sex-chromosomes, which are responsible for sex-determination. Sex-controlled genes may be carried in any chromosome and when, as is usual, they are autosomal, they are transmitted equally by either sex. Their relation to sex is therefore a physiological one.

Many genes totally sex-controlled in their effect are known in other animals, especially butterflies. These give ordinary segregation but in one sex only (which can be either the XY or XX type), though showing autosomal transmission in both. Such inheritance in its complete form is very rare in human genetics, except for the development of the accessory sexual characters, which are not directly due to the sex-genes but to the hormones which are ultimately an outcome of their action: as demonstrated by the tendency to develop incipient beard and moustache in some

[1] This is sometimes called *sex-limited inheritance*, but that term should be avoided as it is too easily confused with sex-linkage.

elderly women. An approach towards total sex-controlled inheritance is provided by the presence of a white forelock, as well
as by frontal baldness. These are both autosomal conditions while
they are dominant in males but recessive, and therefore much
rarer (p. 53), in females.

Partially sex-controlled effects are, however, common in mankind. We are still in doubt whether essential hypertension is
polygenic or due to a single gene with incomplete dominance; on
the other hand, it may be polygenic with one or two of the units
exercising a greater effect than the others. At any rate, it can be
said that the condition affects males far more than females.
Conversely, though the detailed cause of rheumatoid arthritis
is in doubt (p. 106), it certainly has a genetic component and is
commoner in women than in men. It is of course true that many
sexual differences in the incidence of diseases, even in those
determined by a single gene, must be environmental and may
differ greatly from one race to another.

4.2 DOMINANCE MODIFICATION

It will now be clear that the effects of a gene can be varied by
changes in the rest of the hereditary material and, consequently,
that they are susceptible of selection. Indeed the conclusion that
selection can modify the action of a gene, while the gene itself
remains unchanged, is one of the highest evolutionary consequence. Consequently, if the characters so produced are advantageous, they can be improved; if disadvantageous, they can be
minimized. Such effects have now repeatedly been obtained experimentally (Ford, 1940b).

One aspect of this mechanism is that which relates to the evolution of dominance. The possibilities arising from the selective
modification of the effects of genes were first fully appreciated by
R. A. Fisher, and it is to his analysis of them that the concept of
dominance modification is due (Fisher, 1928, 1931). This subject
must first be studied briefly from a general point of view, when its
bearing upon the problems of human genetics can be indicated.
It has been reviewed by Sheppard and Ford (1966) and by

O'Donald (1967). They also answer certain objections to the theory of dominance modification by selection that have been advanced from time to time.

The effects of rare genes are almost wholly limited to the heterozygous phase. This is indeed obvious, since the chances of bringing together two rare heterozygotes, from which occurrence alone a homozygote can arise, must be remote (the actual proportions are discussed by Ford, 1950, see also p. 116 of this book). Disadvantageous genes are constantly subject to counter-selection, so that they must be rare, judged by their frequency in the population at any one time. Nevertheless, the organism must have a wide past experience of them since mutation is a recurrent phenomenon. Therefore, unless such genes are so lethal that their bearers leave no descendants, animals and plants will have some opportunity of modifying their response to them by adjustments of the gene-complex. Such a process will be in the direction of mitigating the effects of disadvantageous genes when in the heterozygous state: any similar adjustment in respect of the homozygote will be relatively minute, owing to its immensely greater rarity. Therefore there will be a constant tendency for selection to modify such disadvantageous effects in the direction of suppressing them in the heterozygote; that is to say, to make them recessive. The converse tendency will, of course, operate upon the advantageous effects of any gene. These will be magnified in the heterozygote so that they will become dominant in expression. Furthermore, the frequency of the gene responsible for them will increase, so providing more material on which selection can act. Consequently the process will be far swifter than the drift towards recessiveness of disadvantageous characters, which depend for their maintenance upon recurrent mutation. It is plain then that if a new gene spreads through the population, its advantageous effects will acquire dominance while its disadvantageous ones will become recessive. For dominance and recessiveness are properties of characters not of genes.

Thus when we study the multiple effects of a single gene, we find that they may have different dominance relationships from one another. On p. 118 a gene evoking three characters is discussed:

one dominant, one recessive and one intermediate in the hetero-zygotes. A further instance of the kind, among many, is provided by the condition already used in giving an account of Mendel's second law (pp. 11–14): red hair in man. This is a recessive and is expressed equally in the two sexes. But the gene responsible for it has another effect; the production of freckles, and these occur in the heterozygotes as well as in the homozygotes. This feature of its action is therefore dominant (Nicholls, 1969). In either geno-type, the degree of freckling is influenced by the environment, being accentuated by exposure to sunlight. The distribution of red hair and freckles in different human races is referred to on p. 194.

Nicholls examined the frequency of these characters in Aus-tralia. In a State Mental Hospital there, 2 individuals had red hair when, compared with the normal population to which the patients belonged, 18 were expected: a heavily significant difference (pp. 194–5).

It should be noticed at this point that the spread of a gene having some advantage is often arrested during its course. This occurs when the characters to which it gives rise reach an optimum frequency in the population owing to a balance of selective agen-cies. The resulting situation is of great practical importance in medical genetics; it will form the subject of the next chapter.

When a character is so disadvantageous that the individuals possessing it contribute little or nothing to posterity, it is apparent that no opportunity for dominance modification arises. In these circumstances, the original condition should be preserved, and we may expect this to be such that two doses of a gene produce a greater effect than one. It is found, indeed, that moderately disadvantageous genes are recessive. Yet a class of genes also exists responsible for extremely disadvantageous effects when heterozy-gous; while the corresponding homozygotes, if obtainable, produce even more serious results so that, in many instances, they do not survive. These are obviously the group in which dominance modification has never taken place. Nor has it in respect of the rare members of a multiple allelomorph series, for these will hardly ever be brought together save in experimental work designed to that end. We find therefore that the heterozygotes between any two rare members of the series are intermediate, while those between each of them and the normal allelomorph are not.

These facts throw much light on genetic action in man. Many of the rare disorders are true recessives. It is noteworthy, however, that this state has not been perfectly attained in some of the more dangerous of them, xeroderma pigmentosum for example (pp. 56–7). On the other hand a large number of conditions, some of a serious kind, are described as 'dominants'; epiloia and Huntington's Chorea may be cited from among them. Yet I have pointed out that this is an assumption in nearly every instance, since the corresponding homozygotes have rarely been studied. The theoretical considerations just outlined suggest that the heterozygous effects of such genes will usually be somewhat intermediate. Some evidence for this conclusion can be obtained in more than one way. First, a very few instances are recorded in which the homozygotes may actually have been observed, and they appear usually to produce the known characters of the condition in a highly exaggerated form. Fraser Roberts (1940) draws attention to a pedigree of a form of 'dominant' brachydactyly in which the only abnormality consists in a shortening of the middle phalanges of the index fingers and toes. An affected individual, being as usual a heterozygote, married his first cousin, the daughter of an affected man. She had probably been affected, though she was dead when the family history was compiled and the point could not be established. The couple had two children. The elder was affected to the same degree as other members of the family, and proved to be a heterozygote (having one abnormal and two normal children). The younger child, however, was grossly abnormal. She had no fingers or toes and her 'whole osseous system was in disorder'. She died within a year, being unable to develop. She was probably a homozygote. Such a condition as this cannot be ranked as a dominant at all.

Further evidence that dangerous disabilities are rarely true dominants is supplied by the variability of their heterozygous expression. This is notable in both the instances just cited, epiloia and Huntington's Chorea.

There is indeed a general tendency for the expression of a gene to be more variable when heterozygous than when homozygous. Fraser Roberts (1940) has stressed that recessive conditions in man are more constant than are heterozygous ones. Indeed I have myself found, in the course of experiments on artificial dominance modification, that the heterozygotes produce a greater range of variability and are more modifiable by selection, than are either of the homozygotes; while of these, the normal is more constant than the mutant (Ford, 1940b). The reasons for this can be stated briefly. There will often be circumstances limiting the possible range of expression of a gene. For instance, increasing amounts of a pigment tend towards saturation in their effect, while the size of an organ cannot indefinitely be increased. This should make the effect of the single-dose condition more susceptible to extraneous influences, whether of changes in the gene-complex or of the external environment, than that due to the action of the two allelomorphs. Further, the normal allelomorph will have been adjusted to produce the most favourable, and therefore rather constant, effects within the range of the internal and external environments to which it is generally exposed.

The degree to which the effects of a gene are variable is described as its *expressivity*. It is this quality with which we deal, for instance, in discussing the fact that the gene producing 'dominant' polydactyly is responsible for characters ranging from well-developed extra digits on both hands and feet to an ill-formed knob in addition to the normal fingers of perhaps a single hand.

Expressivity relates to the degree in which the characters produced by a gene are expressed. On the other hand, the frequency with which a gene produces any effect at all is called its *penetrance*. Most of the genes so far discussed have complete penetrance; they are always responsible for some effect. Occasionally, however, genes with heterozygous manifestation fail to produce any detectable characters: their penetrance is incomplete. For instance, diabetes insipidus is usually due to a single gene with heterozygous expression, a so-called 'dominant' (p. 85). This occasionally fails to cause the disorder, which therefore skips a generation, though it should be transmitted regularly from parent to offspring. Medical practitioners must be especially cautious in advising their patients on the probability of transmitting such genes. With heterozygous expression and full penetrance, unaffected individuals cannot have affected children; but when penetrance is incomplete, occasional exceptions will occur. Should the penetrance of a gene be slight, the condition to which it gives rise will appear sporadically in certain families. In these circumstances, extraneous agencies will often tend to evoke it. There is some evidence that diabetes mellitus is due to a gene with heterozygous expression and about 10 per cent of penetrance.

It has already been pointed out that the effects of homozygotes are usually less variable than are those of heterozygotes: indeed, failures in penetrance do not seem to be encountered very often in recessives. It is possible that the Laurence–Moon–Biedl syndrome, a recessive condition due to a single gene, is not always expressed: that is to say, the penetrance of the gene is not quite complete.

It is, however, clear that the dominance relationships of inherited characters in man are distinctly abnormal. Nor is this to be wondered at considering that they are produced by selection: for the conditions of human society during at least the last two hundred generations have departed widely from those of all wild species. They have in fact been unique, but they have somewhat resembled the situation to which domesticated forms are subject. Even in a primitive human society, individuals must have been valued for and protected by qualities other than bodily adaptations to their environment. Indeed from an early period in the evolution of man, individuals who would have been eliminated from the populations as breeding-units during the pre-human stage must have made considerable contributions to posterity. Such a negation, or in some instances a reversal, of selection will not only fail to produce the normal selective modification leading to dominance, but may break down dominance relationships previously attained. It is in domesticated species that we find the greatest abnormalities in dominance adjustment: a viable dominant black form occurs in the guinea-pig, while the hornless condition in cattle is nearly dominant also. Species whose domestication has been of an unusual kind are those whose dominance adjustment is particularly abnormal. Indeed the peculiarities of domestic poultry in this respect, which at first appeared mere exceptions to Fisher's theory of dominance, have provided by their very irregularities clear proof of its operation (Fisher, 1935, 1938). So too man, whose exposure to selection must evidently be exceptional both in degree and in kind, supplies examples in unusual numbers of a rare condition: that of genes producing heterozygous effects which must be deleterious, but to a slight extent only. It has already been explained that this situation is not normally to be anticipated. Instances are provided by phalangeal synostosis, piebald spotting, congenital cataract and others.

R. A. Fisher was the first to suggest that dominance is produced by selection (1928), and he concluded that this operates upon the gene-complex in the manner already described. Haldane (1930) subsequently proposed an alternative method of dominance modification. He pointed out that genes probably act by producing enzymes, and that these will operate effectively only up to a saturation level, above which further enzyme productions will produce no detectable effect. It has already been pointed out that multiple allelomorphs control a given set of characters quantitatively. Now, assuming that a multiple allelomorph series exists at each locus, Haldane pointed out that dominance modification might arise by selection of different allelomorphic genes in the following way. We may suppose that the homozygote is due to the operation of unnecessarily 'high' members of the series, since selection will favour those allelomorphs whose activity is sufficient to raise enzyme production to the saturation level even in single dose. Rare genes will always be judged by their reaction as heterozygotes, and those will be favoured which even in that phase do not lower enzyme production below the normal saturation level, so as to evoke an imperfect development of the characters concerned. That is to say, Haldane envisages dominance arising through a change in the gene itself, through selection of one rather than of another member of a multiple allelomorph series.

The level of the threshold value up to which an enzyme acts will be determined by the general constitution of the organism and so, in part, by other genes; thus it will be susceptible of modification by selection operating on the gene-complex in the manner proposed by Fisher. It should also be noticed that experimental proof of Fisher's method of dominance modification has now been obtained in a number of instances. It has indeed been possible under experimental conditions to select the heterozygous expression of the characters produced by a given gene until they have become nearly dominant in one line and nearly recessive in the other. In these circumstances, it has been established, by crossing with the normal form, that the gene itself has remained unaltered (Fisher, 1935, 1938; Ford, 1940b).

It has been stressed that the genes of an organism act together to form a gene-complex adjusted by selection to give a favourable result. The chances therefore are exceedingly remote that purely random changes in the genes, such as arise by mutation, shall so fit in with the balanced system already in existence as to promote harmonious working. The probability is indeed overwhelming that the effects of mutations will be disadvantageous, a circumstance to which attention has already been directed. The consideration that mutation is a recurrent phenomenon also contributes to this result. Were organisms living in a constant environment it would seem almost incredible that mutations having advantageous effects should ever now arise. All that did so would long ago have been utilized, and the mutant genes which they produce incorporated into the gene-complex of the organism. But characters which are disadvantageous in one environment may not be so in another, so it is possible that a mutation may occasionally produce useful effects, but we should expect this to be a very rare event; and so it proves. Nearly all the mutations which have taken place in experimental material have been disadvantageous; only in two or three instances has one evoked characters which might in certain circumstances be of use to the organism. Were it not so, were mutations frequently to produce effects of evolutionary value, we might well doubt if the genes which we study represent the type of hereditary material normally employed in organic evolution. The fact that mutations hardly ever give rise to advantageous qualities is in accord with the view that the genes, whose diversity is a product of mutation, provide an adequate basis for the heritable variation upon which selection operates to produce evolutionary change, as well as for the corresponding essential of heritable stability.

4.3 ENVIRONMENTAL VARIATION

Genetic factors interact with the environment to produce the characters for which they are responsible. Part of that environment is provided directly or indirectly by other genes, and that portion of it has now briefly been considered. But the external environment in which the organism lives is an essential element in the production of its characters, all of which are both inherited and acquired; a fact first clearly stressed by Goodrich (1912, 1924). Therefore an alteration in either of these components may lead to variation. Changes in the genetic constitution, due to mutation or recombination, may give rise to *genetic variation*, and changes in the environment in which the genes operate may produce *environmental variation*. An organism judged by its characters is known as a *phenotype*, one judged by its genetic constitution as a *genotype*. Two individuals may be phenotypically similar but genotypically different: as are heterozygotes and their dominant homozygotes. Two men who can taste phenyl-thio-urea are phenotypically similar for this character, but if their constitution is TT and Tt respectively, they are genotypically different. On the other hand, two individuals may be genotypically similar but phenotypically different, owing to the operation of environmental variation, and it is this phenomenon which we must now briefly study.

Indeed as often pointed out, the environment is, in all organisms, as important as heredity in producing the characters of the individual but environmental *variation* is particularly difficult to examine in the mammalia, since they are protected against it to a high degree. The reason for such protection is not difficult to find. Every organism must have an optimum environment, one which suits it best. Obviously it is to the advantage of all individuals to live as near to their optimum conditions as possible. This they may do by migration, by adjusting their habits and ecology to changing conditions, or by manufacturing as far as they can a satisfactory environment of their own which they carry with them and in which they permanently live. This latter technique has been perfected by the mammalia in a remarkable way. In lower forms it is possible to study the bodily effects of altering the more obvious components of the environment, such as temperature and humidity. Yet the thermostatic regulation of the mammalia is such that a departure of even a few degrees from their optimum temperature, which they themselves maintain, is fatal to them. Similarly, changes in humidity can have but a limited effect upon them, while the constitution of their body-fluids is regulated with great exactitude. It is plain then that mammals live in an extremely constant environment, one which approximates to their optimum, so that they are only subject to environmental variation to a small extent.

We are provided in lower forms with innumerable examples of the fact that the genes produce different effects in different environments, so that there is little chance of falling into the fallacy of regarding them as the inherited basis of particular characters. In mammals, where the amount of environmental variation is so much less, this mistake is more readily made. It can easily be shown in the majority of organisms that changes in temperature, in humidity, or in other aspects of the external environment, will modify the action of the genes. Yet there are instances in which such effects can be detected within the small environmental range which the mammals can experience. This is true even for a condition controlled so accurately as temperature. The point is sufficiently striking to merit a brief description.

The 'Himalayan' rabbit is a white form with blackish extremities. It is inherited as a simple recessive, being due to the operation of a gene c^h, which is a member of the albino series of multiple allelomorphs (p. 24). Himalayan rabbits are born wholly white, since the gene is one which prevents pigment formation at the mammalian constant temperature. A few degrees lower, however, its action is quite different and leads to melanin production. Consequently, the extremities darken, for they lose heat more rapidly than the rest of the body and fail to attain the true body-temperature. Thus the exposed parts, the tip of the nose, ears, tail and the feet are blackish in adult Himalayan rabbits. It can be shown experimentally that the Himalayan character is not due to the operation of a gene controlling pattern, but to one which produces different effects at different temperatures. If an area on the back of Himalayan rabbits be shaved, the hair which grows again will be white if the animal be kept warm and black if it has been kept in cold conditions. Several other genes interacting with temperature in a similar way have been studied in the mammalia. Examples are provided by those responsible for two of the three chinchilla phases in the rabbit, and for the Siamese cat.

No such striking environmental effects have so far been detected in man, and our knowledge of environmental variation in human genetics is small. It must be sought rather in less obvious directions. One of these is the effect of maternal age upon the penetrance of the genes possessed by the offspring. It is known that a number of conditions, some of them in part genetic, are expressed with greater frequency in children as maternal age advances. An instance of these is, as already explained, provided by mongolian idiocy (p. 64.)

The allergies seem due to various environmental stimuli operating in the presence of an autosomal gene; its symptoms described as 'dominant'. However, work done by Wiener et al. (1936), suggests that its effects develop before puberty when homozygous, the heterozygotes giving rise to allergy starting after puberty, or else to normal 'carriers'. The condition is partially sex-controlled, since about twice as many males as females are affected; the sexual difference presumably being restricted to the heterozygotes. Allergic disease may assume numerous forms: infantile eczema, Besnier's prurigo, lichen urticatus, urticaria, light- or cold-sensitized skin, food-allergy, asthma, hay-fever, migraine and others; including angioneurotic oedema, of which a distinct non-allergic form also exists, inherited as a simple 'dominant'. Sometimes, however, a given form of allergy is inherited in a family, or in a small group within a family in which various types appear. Actually, a single gene is responsible for all, but other genes determine which tissues are particularly liable to become sensitized in its presence.

The influence of the environment in evoking the effects of genes is apparent in a number of diseases. As is well known, rickets is due to a deficiency of vitamin D. This may arise from an insufficiency of those articles of diet containing it, such as fish-oil, eggs, butter, milk and cheese, or to under-exposure to sunshine; for ultra-violet light leads to the production of vitamin D in the skin from pro-vitamins such as ergosterol. The disease is common among children of the poorer classes brought up in the sunless conditions of big cities. But it arises so much more easily in some constitutions than in others that before the discovery of vitamins it was classified simply as an inherited defect. Among slum populations exposed to an apparently similar environment, it will occur frequently in some families but not at all in others; while children who develop rickets may have healthy brothers or sisters, though there may be no obvious difference in their upbringing. There is in fact fairly strong evidence that, in some cases at least, susceptibility to the condition is due to the action of a single gene with heterozygous expression. When both vitamin D and a sufficiency of sunshine are lacking, this would ensure the inheritance of rickets upon simple Mendelian lines, though the segregation of the gene would be completely obscured in a family supplied with a well-balanced diet. This is clearly a parallel situation to that described in Chapter 1 in which the production of yellow fat in rabbits is unifactorial and recessive, giving rise to strict Mendelian segregation but only in animals supplied with green food. Without this article of diet, the presence of the gene is undetectable.

As already mentioned, susceptibility to one aspect of the environment, that is to say infection, is often inherited (see also pp. 163, 196). Grüneberg (1934) showed that a single gene with heterozygous expression is responsible for an inflammatory condition of the accessory nasal passages. The penetrance of the gene is incomplete, since its action depends upon exposure to infection. It can often be stated in general terms that some constitutions are more liable to infection than others, though the genetic control involved may be entirely unknown. Doubtless it must often be multi-factorial. When fraternal twins (pp. 186-7) are born of a syphilitic mother, instances are known in which one only has contracted the disease, though their opportunities for infection must have been equal and great. Such a difference as this must almost certainly be genetic.

The genetic influence in infectious disease is of course more difficult to assess with advancing age, since the hereditary constitution becomes one only out of many agents predisposing to immunity or the reverse. The medical history of each individual and his past and present habits of life, in addition to his genetic outfit, are of great and obvious importance in determining resistance to infection.

4.4 CANCER

The interaction of heredity and environment is also evident in cancer. There can be no doubt that one of the chief agents predisposing to this disease is long continued irritation at a point; yet such a stimulus produces a malignant change in some individuals but not in others, and with unequal frequency in different parts of the body: thus the mucous membrane of the nose is much subject to irritation but it seldom becomes cancerous. These facts indicate that susceptibility to cancer is to some degree genetically controlled, a suggestion which has now been verified by decisive evidence. Some of this can be obtained from the *British Journal of Cancer* (1948, Vol. II, part 2), in which are published the papers presented to the International Symposium on the Genetics of Cancer organized jointly by the Genetical Society of Great Britain and the British Empire Cancer Campaign. Here it will be possible only to give a very brief survey of the subject.

One of the great difficulties in studying cancer is the extremely diverse nature of this disease, and every gradation exists in its genetic control. At one end of the series is the tendency, determined on a multifactorial basis, to raise somewhat the susceptibility of the body as a whole, or of particular organs, to malignant change. The other end is provided by unifactorial conditions which so interact with unavoidable stimuli as to produce cancer segregating in certain families in simple Mendelian proportions: xeroderma pigmentosum (pp. 56-7) and the gene heterozygous in expression which produces pre-cancerous polypi of the colon and

94

rectum, are examples of this kind. So is the situation reported by Howel Evans *et al.* (1958) who found two, probably related, Liverpool families in which tylosis is associated with cancer of the oesophagus. The skin condition is transmitted as a simple heterozygous defect, I am not aware of evidence to show that it is dominant (see p. 85). In the two pedigrees combined, 18 of those suffering from it have developed the malignant growth, while no non-tylotic member of either family has done so. There is no indication that the epithelium of the oesophagus is abnormal in those suffering from the skin defect. Though it is likely that we are here dealing with distinct effects of the same gene, the possibility of very close linkage between two mutant loci cannot be excluded.

Tylosis consists in a thickening of the skin of the palms of the hands and the soles of the feet, occasionally of the latter only, and normally has no connection with oesophageal cancer. It exists in two forms, each controlled as heterozygous conditions. The commoner arises during the first year of life and seldom causes discomfort. The rarer, which is that found in the two Liverpool families in which it is associated with malignancy, which normally it is not, appears later, between the ages of 5 and 15 years. Moreover, it often gives rise to painful skin fissures which frequently become infected.

The preventive aspect of the situation is evident. Successful removal of the malignant growth is generally possible when occurring as it usually does, near the lower end of the oesophagus; but only during the early stages of its existence, when symptoms may be slight or absent. The presence of tylosis, however, indicates the individuals in those particular families who are at risk and, provided they show themselves for examination with sufficient frequency, it may be possible to operate early enough to save them. The growth has started in 95 per cent of all those tylotic cases in these families who attain the age of 65, while it has been recorded as early as 35.

Not only can distinct individual genes produce similar effects (p. 33) but corresponding multifactorial characters can be built up genetically in distinct ways. Thus the extensive data of the

Danish Cancer Registry demonstrate the reality of hereditary tendencies to raise the susceptibility of the whole body, or of individual tissues, to cancerous degeneration. They indicate moreover that a marked family history of cancer of the breast is associated with an unusually early onset and with a somewhat higher frequency of cancer in other parts of the body (Kemp, 1948). On the other hand, the data of Smithers (1948), while confirming the genetic predisposition to breast cancer, afford no indication that this is associated with a higher frequency of this disease at other sites or with an earlier onset where there is a family history of it. These results, which agree in ascribing a genetic basis to cancer of the breast, must not necessarily be held contradictory in other respects: the reactions of the genes concerned may not necessarily be identical in the environment, genetic or otherwise, provided by the Danish and British populations from which the evidence was respectively obtained.

A special warning must here be given against drawing too close a parallel between genetic situations in human cancer and those ascertained by means of experimental studies on other animals. Valuable as they may be for the general theory of the disease, they are not necessarily applicable in detail to parallel forms of cancer found in man.

It is now known that certain virus infections can fuse single cells together in man and other animals: a situation which can be used experimentally to combine different cell-types, whether from the same or distinct species. This property has been employed in the study of cancer. Harris (1971) fused highly malignant mouse cells with non-malignant ones. They have then been transferred into individuals which had been irradiated in order to prevent them from rejecting the transplant. Instead of the tumour developing immediately in all the mice, as it does when malignant cells alone are injected, cancer did not arise as long as the hybrid nuclei retained the full double set of chromosomes. Subsequently, as they began to suffer from chromosome loss, these cells started to become malignant, so that selective chromosome loss may be a factor in causing some forms of cancer. It is highly curious that the normal healthy cells seem at first to check malignancy; a situation so different from that found when a cancer is fully established; it is one that deserves much fuller study.

Segregation and somatic mutation, sometimes perhaps of the rare genes of the cytoplasm in addition to the normal nuclear type carried by the chromosomes, must both be agents in producing cancer. This, indeed, is strongly suggested by the discovery of Demerec (1948) that some carcinogenic substances produce mutations while others do not. It should be noticed, however, that even those malignant changes which are mutational in origin may yet be genetically determined, for the mutation-rate is under genetic control. Some genes, as well as cytological abnormalities, pp. 43 and 64, may have a physiological effect which tends to promote internal environments predisposing to cancer. Thus evidence has been obtained by Aird, Bentall and Fraser Roberts (1953) which indicates that cancer of the stomach is especially associated with blood group A of the ABO series (pp. 162–3).

The relationship between lung cancer and smoking has been handled very clearly by Clarke (1964), whose account of this subject should be consulted. A few salient points only can be dealt with here.

In this matter, there appears to be no doubt that the chief danger arises from addiction to cigarettes rather than to other methods of smoking, and there are of course several factors which distinguish that aspect of the habit, though it is not known which of them is important from this point of view. Thus the tobacco used in cigarettes is not the same as that employed for pipes and cigars; it is differently cut and a different temperature is probably reached as it burns while, of course, in cigarettes, paper is smoked as well as tobacco.

A well constructed and large-scale analysis of the effects both of cigarette smoking and of heredity in producing lung cancer was carried out by Tokuhata and Lilienfeld (1963). They used 270 patients and their first-degree relatives carefully matched with 270 controls and their first-degree relatives; amounting to 3,700 individuals. To provide a comparative basis, they attributed a risk of 1·0 to a non-smoker with no family history of the disease. For a non-smoking relative of a lung cancer patient, the risk becomes 3·96. It rises to 5·25 for cigarette smokers without affected near relatives, and reaches 13·64 for cigarette smokers with a relation who has developed lung cancer. Thus the result of this survey has demonstrated that cigarette smoking and a hereditary tendency can both be involved in causing the neoplasm.

Selection is powerless to control those harmful qualities which, though genetically determined, arise after the normal age of reproduction. However, one could at least in theory imagine conscious selection acting in man, and easily on domestic animals, to reduce the frequency of diseases in later life. For instance, if there were a tendency to avoid marriage into families whose elderly members were known to have a high incidence of cancer.

Selection is also ineffectual in adjusting the population against very rare defects, not because there is any minimum threshold for the amount of variability which selection can influence, but because the process of adjustment in relation to very rare variation may be so slow that what is attained in the earlier stages may be inappropriate in the later ones owing to other and more rapid evolutionary changes. I have pointed out (Ford, 1949) that malignancy of epithelial cells appears to be sufficiently common for selection to have buffered the body against it with considerable efficiency, for carcinomas are quite rare before the age of forty. It is suggestive also that the rare connective-tissue neoplasms occur as commonly in early as in later life. It should be noticed that xeroderma pigmentosum, in which epitheliomas of the face and hands appear even in childhood, is no exception to the selective adjustment of the human population to cancer, since this is a unifactorial condition which has been pressed into nearly complete recessiveness relative to its pathological aspect.

Similar genetic tendencies in the occurrence of cancer are to be observed in other animals. Thus some species, and varieties within the same species, are much more susceptible than others to malignant diseases. These are relatively frequent in domestic fowls (79 out of a sample of 880, 9 per cent) and rare in cattle (131 out of 47,362, 0·2 per cent). It is a striking fact that about 80 per cent of grey horses of both sexes develop malignant melanomas of the skin around the anus on reaching old age. The frequency of all cancers in horses of other colours is given as 5 per cent. This is calculated on individuals with a much younger mean age, but there can be no doubt of the extraordinary frequency of melanomas in the grey animals (Raven, 1950).

The genetic component in the development of cancer should be of importance in medical practice, for the possibility of this disease should especially be kept in mind in treating middle-aged or elderly members of a family in which several cases of it have occurred. Moreover, early diagnosis is the pre-eminent necessity if cancer is to be treated with reasonable hope of success, and anything which contributes to its detection before dissemination by metastases is of the utmost value. I have myself encountered two instances in which patients sought medical advice for vague abdominal discomfort several months before an exploratory operation disclosed a carcinoma which was then too advanced for operation. In both of them there proved to be a marked family history of cancer. Had that fact been taken into account, which it was not, it is conceivable that suspicion might have been aroused earlier, at a time when the prognosis was not yet hopeless.

A form of cancer which raises several unusual features must be mentioned here. This is the chorioncarcinoma. The implantation of the mammalian embryo bears some general resemblance to the behaviour of a neoplasm. In both, the cells are relatively undifferentiated, and both erode and destroy the normal tissues. It would seem that we have here something more than a facile comparison. For though in the one situation the destructive process continues without limit while in the other it is normally brought to a halt, yet occasionally the mechanism which checks the invasion of the placental cells fails, early or late in pregnancy, and they then actually form a cancerous growth. It will be noticed that the chorioncarcinoma is a true parasite for it is derived from another individual, the foetus not the mother.

It looks as if a genetic component is operating here, for the condition occurs very unequally among the races of mankind. It is alarmingly common among Chinese women. Its frequency among them has been assessed only roughly; but a value of 1 in 1,000 who have had a conception seems to be regarded as a reasonable figure. Among western Europeans, however, the disease is exceedingly rare.

Bagshawe (1971) finds that in England the highest risk of chorioncarcinoma is experienced by women of blood group A married to group O men, and the least by group A women married to group A men; the difference being in the ratio of 10·4:1. There is indeed a parallel with Rh immunization here (Clarke, 1971*b*) which is more likely when the blood group of mother and foetus are the same (O and O) than when compatible though different (A mother, O baby). Spontaneous regression of the neoplasm occasionally occurs after the evacuation of a hydatidiform mole, and it does so most commonly in those mated to their own ABO phenotype. Bagshawe also shows that in group AB patients chorioncarcinoma tends to be rapidly progressive and to respond badly to chemotherapy.

These serological distinctions may relate to the fact that metastases of the chorioncarcinoma are in some cases slow to start and rather restricted, in others rapid and widespread. Professor C. A. Clarke (1964, pp. 494–6) makes a perceptive suggestion in this matter. He points out that while female mouse tumours implanted into female mice are generally accepted, male tumours are not; the female making an antibody against an antigen due to a Y-borne gene. This may well throw light upon the late compared with the early and rapid metastases in chorioncarcinoma, depending on whether the malignant cells are derived from a male or a female foetus.

Indeed one of the more hopeful possibilities in attacking the cancer problem is that of inducing the body to reject the malignant cells. There are signs, in addition to the behaviour of the chorioncarcinoma, that this approach may be a fruitful one (Clarke, 1971*b*). Drugs are used to suppress allergic reaction against a kidney graft, which is otherwise rejected, and there is a strong indication that the frequency of lymphoma and lymphosarcoma is high in those who undergo this treatment.

Fisherman (1960) demonstrated a significant association between cancer and allergy, using 1,185 cancer cases and 294 controls. He found that 3·2 per cent of the cancer patients had allergic symptoms while these were present in 12·9 of the controls. Mackay (1966) also made a carefully planned study which showed that allergic disorders were significantly less common in association with malignancy than in his control group; the correlation was limited to women, which it was not in Fisherman's series: a distinction between the two groups which remains unexplained. Mackay further makes the supporting comment that, in general, allergic symptoms tend to decrease and the frequency of cancer to increase with advancing age; yet this is a consideration which can also be viewed in another way.

The integration of cells in such a manner that they can build up a harmoniously adjusted 'population', and their successful differentiation into a variety of types adapted to perform diverse functions, is likely to have posed problems of great difficulty to organisms from the inception of the Metazoa. As in all situations requiring specialization, this must, particularly in the earlier stages, have constantly broken down, leading in this case to attacks by dedifferentiated cells upon their neighbours. The cancer problem must therefore go back to a remote epoch in evolution; consequently we need not be surprised to find, as we do, its manifestation in plants, and in lower and higher animals. Evidently the tendency for cells to become malignant must always have been strongly opposed by selection.

Bearing these considerations in mind, we may say that the human race had largely cured itself of cancer up to that stupendous economic revolution when man passed from food-gathering to agriculture and stock-raising: from Mesolithic to Neolithic conditions, the latter first typified for Britain by the Windmill Hill Culture, a little after 2300 B.C. For in the environment which persisted up to that time, a life-span of forty years seems to have been about a maximum, seldom attained. In such a population, cancer would be quite a rarity. The far more stable and favourable conditions experienced by Neolithic people would allow survival well beyond the previous limit, into an age-group which selection had been given no opportunity to influence: one indeed which it could not influence beyond the period of reproduction (Ford, 1949). The barrier against certain diseases would thus be lowered in later life; most of all against so fundamental a condition as cancer. On the other hand, there would then be extended possibilities for immunity against infections and, where this could be done, to adjustments: perhaps against those allergies which start in childhood, but not against the rarer forms which first manifest themselves in middle age.

It might be thought beyond the power of selection to adapt cells to withstand the physical effects of penetrating radiation. Yet there is good evidence to show that some resistance to them has in fact been achieved and is in part genetic. This has been obtained both in mice (Grahn, 1958) and in *Drosophila* (Parsons *et al.*, 1969, and by others) and probably therefore it has an extremely wide application in organisms. Thus the essential part taken by selection in checking cancer must extend to increasing resistance to such short-wave radiation as is normally encountered. This comprises background radiation, the biological effects of which are little known, and cosmic rays. These latter reach the earth as the result of radio-activity in stars. To quote a highly discerning passage by Clarke (1964, p. 267): 'Cosmic rays may be of great importance in space travel, but their effect on those who remain earthbound is unknown. However, *a priori* it is reasonable to assume that there will have been selection for individuals capable of withstanding any deleterious influences that the rays may have exerted.'

Cosmic rays, the components of which are of several kinds, including protons and neutrons, are largely filtered out by the atmosphere. Consequently their intensity increases with altitude; it is six times greater at 15,000 feet than at sea-level. Clarke's wise comment should therefore indeed give pause to those who encourage space travel.

4.5 GENETIC PHYSIOLOGY AND PHARMACOGENETICS

The physiological steps by which the genes produce the characters for which they are responsible have now been studied in many instances. This has involved a combination of biochemistry, genetics and physiological experimentation; one which has proved very fruitful. The whole subject, including its extension into pharmacogenetics (pp. 107–9), has been skilfully surveyed by Clarke (1964, 1969, 1971a) who has himself contributed greatly to it. His books should be consulted by those who require a detailed introduction to this type of work, together with the necessary references. For here, owing to lack of space and concentration upon principles, it is impossible to do more than indicate its scope, illustrated by means of a few examples.

In the first place, however, it should be remarked that Ford and Huxley (1927), working from 1923 onwards in the early days of the subject, established that genes control rates of processes in the body and the time of their onset. There can be no doubt of the general importance of this concept, and its application to man may briefly be considered.

It is a striking fact that the newly born human baby possesses certain features usually found only in the foetus of other mammals; the cranial flexure, for instance. This circumstance has been somewhat over-stressed by Bolk (1926), who held that a process of 'foetalization' has been largely responsible for human evolution. This seems to be an exaggeration, yet it is clear that an alteration in timing, such as the delay in closure of the cranial sutures and the lengthening of the baby phase, has played an important part in the establishment of the definitely human characters. Doubtless they have been produced by selection favouring genes which slow down rates of growth and development, differences in which are often to be observed among multiple embryos, even in the remarkably constant environment of the mammalian uterus. These must be largely genetic, and represent the type of variation available for such selection. Where there is competition between embryos and between young, those individuals will be strongly favoured which possess genes speeding up growth and development. The opposite has been true in human evolution which, depending upon intellectual advance, has required a slowing down of development and prolongation of the baby phase and the period of learning; a process that has been possible only because mankind usually produces but one child at a birth (and see p. 188).

So far no reference has been made in this book to eye-colour, although it is one of the more obvious of human characters. I have avoided the subject on account of the unsatisfactory state of our knowledge respecting its genetic control. The view has long been held that the distinction between the presence of brown pigment in the eye and its absence, which gives rise to blue or grey shades, is unifactorial, brown being dominant. This is correct, but it is undoubtedly an over-simplification. The point to which special attention must here be drawn is the circumstance that all human eyes save the extreme albino (in which black pigment is absent from the tapetum) are blue, due to a scattering of light by connective tissue, unless brown pigment be deposited in the iris. However, when this occurs, it does not take place until after birth, so that the eyes of infants are usually blue, whether they are destined to remain so or not. It appears highly probable, moreover, that variations in the rate of deposition of the brown pigment and the time of its appearance are responsible for some of the differences in human eye-colour, and that these are genetically controlled.

We may now turn to the more physiological aspects of human genetics. These may in the first place be illustrated by the metabolism of phenylalanine (Fig. 11). This takes two routes. One leads to the production of melanin from tyrosin, a step prevented in those who lack tyrosinase: a condition which therefore produces albinism. This, as is well known, is recessive and unifactorial.

The chemical pathway from phenylalanine leads also to the formation of phenylpyruvic acid (Fig. 11), and thence to hydroxy-phenylpyruvic acid. The latter step is prevented by a gene recessive in effect responsible for the disease phenylketonuria. The patients excrete phenylpyruvic acid in the urine. Moreover, the same gene is responsible also for what appears a very distinct effect, extreme mental deficiency. Further along the same metabolic series, an enzyme enables man to oxidize homogentistic acid (Fig. 11). The absence of this enzyme, due to a single factor-pair recessive in effect, leads to alkaptonuria. This tends to produce arthritis, and usually to blackening of the bones and cartilages.

As a further example, we may take another metabolic series; one in which glucuronic acid gives rise to hexose phosphate, and it does so to some extent by way of the 5-carbon sugar xylulose. This step is prevented in certain people, almost entirely Jews: a condition controlled as a simple recessive. In such cases, the xylulose, which can be detected because it is a strong reducing agent, appears in the urine, producing 'pentosuria' which may be confused with diabetes.

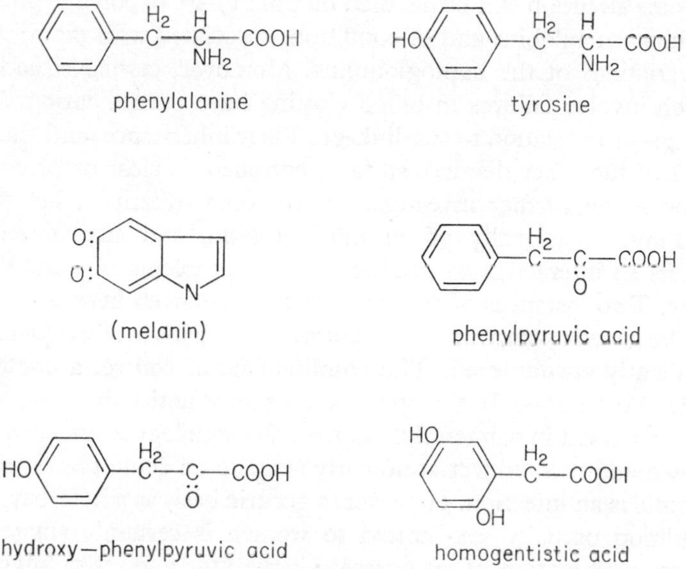

phenylalanine

tyrosine

(melanin)

phenylpyruvic acid

hydroxy—phenylpyruvic acid

homogentistic acid

FIG. II.

A considerable number of conditions analogous with those just mentioned have been analysed: that is to say, a metabolic series altered or stopped at some point owing to the genetically controlled absence of a necessary enzyme.

It is evident that genes affecting the development or action of the endocrine organs are capable of having profound physiological effects. The association between thyroid disease and the ability to taste phenyl-thio-urea (p. 124) is of this type. A further instance is provided by diabetes insipidus which may be either genetic or 'environmental' in origin. The chief symptoms are intense thirst and polyuria. It is due to a single factor, heterozygous in expression, which interferes with the action of the posterior lobe of the pituitary body. Of this the disease is the direct consequence; so much so, that it can be produced also by a lesion of the posterior lobe, resulting from syphilitic meningitis or other causes. In such a situation as this, the interaction of heredity and environment is rather clearly revealed.

Some attention will be devoted on pp. 117–21 to polymorphisms of the haemoglobins and of conditions associated with them, such as variations of the haptoglobulins. Moreover, certain situations which involve failures in blood clotting have been mentioned on pp. 49–52 in relation to sex-linkage. Their inheritance, and that of most of the other diseases so far mentioned, is clear or promises to be so on further investigation. In some situations, however, heredity, occasionally of an unusual kind, and environmental factors so interact as to produce genetic problems of some difficulty. Two instances of this may usefully be given here.

The genetic component of rheumatoid arthritis is of a type fairly frequently encountered. The condition is, of course, a common and important one. In the first place, we must notice that it is much more frequent in women than in men. Its aetiology is not certainly known and it is subject at an early stage to an acute phase which resembles an infection. However, a genetic basis as a heterozygous condition partially sex-limited to women is certainly suggested by an examination of pedigrees. Clarke (1964, p. 185) suggests that this disease may be partly infectious and partly allergic, the latter aspect providing perhaps the hereditary component. This may be sex-controlled. Evidently there are unsolved complexities here. On the other hand, primary osteoarthritis is rather clearly a heterozygous condition, also with marked sex-limitation to women. The secondary type can arise as a result of several conditions: among them, alkaptonuria (p. 104), osteoarthropathy, hereditary dislocation of the hip (p. 195) and osteochondrodystrophy.

Problems are also associated with the genetics of Leber's syndrome (hereditary optic atrophy), in which these are, at least superficially, of a more abnormal type. In this disease, it seems that men are affected and women are carriers, but normal sex-linkage is excluded because the defect is not transmitted by the males who suffer from it. It has therefore been suggested that it is due to genes carried in the cytoplasm. Affected women have been encountered occasionally, but this in itself poses no great difficulties of interpretation. Though perhaps unlikely, it is not impossible that this syndrome really is cytoplasmically determined. Cytoplasmic inheritance has indeed repeatedly been detected as an abnormal condition; especially in plants, in which plastid inheritance provides an example of it. The important point to notice here is that it must of necessity be rare, as indicated by two fundamental considerations. First, we know that the two sexes contribute in approximate equality to the heredity of the offspring (pp. 25–6), whereas practically all the cytoplasm is transmitted by the female. Secondly, genes in the cytoplasm cannot undergo Mendelian segregation; depending, as it does, upon the special features and distribution of the chromosomes. Consequently the variation which they promote can only result from mutation. Yet if this be frequent enough to supply the heritable variability upon which natural selection can operate, it is too frequent to provide the genetic stability which is just as necessary as genetic variation (pp. 27, 32–3).

Turning now to an allied subject, pharmacogenetics has been defined as the study of genetically determined variations revealed by the effects of drugs. The subject has been greatly illuminated by C. A. Clarke and his colleagues at Liverpool. It may be illustrated by three examples.

Hydrogen peroxide normally froths when dropped on to a raw surface, and the blood does not change colour, because the peroxide is degraded by an enzyme, catalase, in the tissues, preventing the oxidation of the haemoglobin. In acatalasia patients, that enzyme is lacking, a condition inherited as a single recessive. Thus hydrogen peroxide does not froth on contact with their tissues, which it turns brown. Those so affected suffer from ulcerating lesions of the gums leading to loss of the teeth.

The causation of such lesions is now clear. Haemolytic strepto-cocci produce hydrogen peroxide; and in those genetically lacking the catalase, the haemoglobin reaching an infected area is oxidized, so producing necrosis owing to local absence of oxygen. Once all the teeth have been lost or removed, the symptoms generally disappear.

Patients occasionally cease to breathe when given suxamethonium in order to promote relaxation of their muscles during anaesthesia. Respiration can, however, usually be restored in them. Its interruption is due to the presence in the affected individuals and some of their relatives of an exceptional form of the enzyme cholinesterase, by which the drug is ordinarily broken down (Lehmann and Ryan, 1956).

The action of this enzyme can be assessed quantitatively by an anti-cholinesterase preparation, dibucaine. This inhibits it by about 79 per cent in normal individuals while in those who react unfavourably to suxamethonium, who amount to somewhere between one person in 1,000 and one in 5,000, the corresponding figure is about 16 per cent. There is also a third group with a frequency of approximately 3 per cent of the population, in which the percentage inhibited is about 62 per cent and these represent the heterozygotes. However, a second abnormal gene has now been discovered. Possession of one normal allele together with either of the abnormal ones ensures a safe response to the drug, while the double heterozygotes, as well as the more recently discovered type of homozygotes, are affected unfavourably to a moderate degree (Lehmann et al., 1963).

Haemolytic anaemia, leading to darkening of the urine, occurred as an occasional complication as soon as primaquine was introduced as an anti-malarial drug. It is found that one of the enzyme-systems concerned with the metabolism of glucose in the erythrocytes is glucose-6-phosphate-dehydrogenase (G6PD). This proves to be diminished in quantity in primaquine-sensitive individuals, leading to destruction of the red cells owing to interference with their glucose metabolism. The deficiency of G6PD is inherited

THE ACTION OF THE GENES

as a sex-linked 'dominant' (and see p. 54). Consequently, males fall clearly into two groups, those who are sensitive to primaquine (that is, they develop the anaemia) and those who are not, while intermediates are found among females. The condition is polymorphic (see Chapter 5) in many areas: 15 per cent of men and 2 per cent of women proved sensitive in a Negro population tested for this distinction.

It has been found that most populations consist of two main classes in respect of their reaction to the drug isoniazid, introduced in 1952 for tuberculosis therapy. Individuals may be either rapid or slow inactivators of this substance, the distinction being clear six hours after oral administration.

This is a polymorphism (p. 110). Slow inactivation is inherited as an autosomal recessive, the two types being in approximate equality in both European and in Negro populations. However, rapid inactivators are much the commoner in the Japanese (they amount to more than 90 per cent) and among Eskimos.

There is a much greater range of variation among the slow type due, it has been established, to the operation of modifying genes. Moreover, the heterozygotes are distinguishable, since the isoniazid concentration in their plasma is higher than that of the dominant homozygotes: due, probably, to heterozygous advantage. In what this consists is not known, but rapid inactivation appears to be particularly important in the Far East and in the Arctic. It is noteworthy that the slow inactivators are the more likely to develop polyneuritis from the use of this drug.

These findings provide an instance of a general proposition now emerging from the study of pharmacogenetics. That is to say, it must be recognized that different individuals may have widely dissimilar responses to certain drugs, the dosages of which need to be controlled accordingly. Also that a corresponding situation exists among the test animals used in drug assay: a consideration unlikely to surprise geneticists but of obvious medical importance.

CHAPTER 5

Polymorphism

5.1 INTRODUCTION

Polymorphism is defined as the occurrence together in the same habitat of two or more discontinuous forms, or 'morphs', of a species in such proportions that the rarest of them cannot be maintained merely by recurrent mutation (Ford, 1940*a*). It will be helpful to explain and expand this definition, and then to consider its implications in some detail. In its earlier stages this account may appear but little related to the practical needs of human genetics. It will later become apparent that this impression is erroneous. Some knowledge of the theory of polymorphism is essential for a clear understanding of the blood groups and kindred phenomena. It will be seen, moreover, that such theoretical analysis provides the only basis on which certain conclusions of importance can be reached.

At the outset, it will be noticed that the definition of polymorphism excludes the following types of variation. (1) Geographical races, such as the white, mongolian and negroid types of man. These originated, and are normally maintained, in isolation from one another, while crosses between them show that they are multifactorial (see the next heading) so that a mixed population of Whites and Negroes, as in some parts of the U.S.A., is not polymorphic. It should be stressed, however, that the occurrence of polymorphism in one district and its absence or different nature in another, may be an important attribute of distinct communities. (2) 'Continuous variation' under multifactorial (or environmental) control, as in human height. This is brought about by the cumulative effects of segregation taking place at many loci. It is not the effect of one or a few pairs of 'switch genes' maintained in the population by selection and determining alternative forms.

POLYMORPHISM

(3) The segregation of rare recessives, albinism for example or rare heterozygous conditions, such as Huntington's Chorea. These are eliminated by selection and maintained only by mutation. They cannot give rise to polymorphism, since the spread of even very slightly disadvantageous genes is checked in its earliest stages (Fisher, 1930a).

The adaptations controlled by a polymorphism are often complex, involving widely different characters due to the action of distinct genes which, if they are to co-operate, must be brought together as a super-gene (pp. 28–33), the members of which will tend to segregate together as a single unit. That indeed is a most important aspect of this fundamental type of variation: the one which allows the organism to incorporate otherwise, perhaps, unusable genes into a balanced and effective group (Darlington, 1971).

In addition to genes which are eliminated by selection, it is necessary to consider the possibility that polymorphism may sometimes be due to those which are neutral as regards their survival value. However, Fisher (1930b) has demonstrated that, in order to produce effective neutrality of this kind, the balance of advantages between contrasted allelomorphs must be extraordinarily exact, so that such genes must be very rare. In addition, without the help of selection their spread through the population is excessively slow. Indeed the number of individuals which possess a gene of approximately neutral survival value cannot greatly exceed the number of generations since its occurrence, if it be derived from a single mutation (Fisher, 1930a). Furthermore, with particulate inheritance, mutation is so rare (pp. 66–8) that its recurrent nature cannot hasten the process very materially. Therefore the spread of such genes will require a period of time too great to fall within the scale at least of minor evolutionary trends. Consequently, since environments are far from constant, we can be sure that a gene neutral compared with its allelomorph will have made little progress before the accurate balance needed for such neutrality will have been upset.

Polymorphism cannot normally be maintained environmentally (Ford, 1965, pp. 11–12). Therefore, setting this aside, and bearing in mind the considerations just outlined, we may be fairly confident that when a form unifactorially controlled occupies even 2 or 3 per cent only of a given population it must possess some advantage. In these circumstances, it may either be in actual process of spreading, or else it may be maintained in a constant proportion by a balance of selective agencies.

These two alternatives, respectively named *transient* and *balanced polymorphism* (Ford, 1940a), each represent conditions of outstanding interest. In the transient type, a form previously rare spreads through the population, which becomes polymorphic during the process. Such an occurrence is probably always due to a change in the environment or in the gene-complex, which gives some value to a character previously disadvantageous. It is perhaps unfortunate that this must be ranked as polymorphism at all, and that the term cannot be reserved for the distinct class of phenomena in which stable ratios are involved. Yet the frequent difficulty of deciding in what instances stability is in fact attained has made the inclusion of the transient situation a practical necessity. To take an example: black forms of many species of moths have appeared in manufacturing districts and have largely supplanted the normal pale types there during the last fifty or sixty years (see Ford, 1937, 1972). While this change was in progress, these populations were in a state of transient polymorphism. Yet this has nowhere led to the condition in which the former normal allele survives merely as a rare mutant, but the situation has become stabilized by a balance of selective advantages and disadvantages. Thus the dominant black form of the moth *Biston betularia* first appeared in Manchester in 1848. Though the insect has but one generation in a year, 98 per cent of the population there was black by 1895; but the further spread of the condition was then arrested owing to the evolution of heterozygous advantage (pp. 113–14), after which the fitness of the pale form *cc* was about half, and of the homozygous blacks *CC* about 92 per cent, that of the heterozygous blacks *Cc*.

A superficially similar situation in man seems to be provided by the occurrence of an exceptional type of pseudocholinesterase in 30 per cent of the inhabitants of Tristan da Cunha. These people, numbering 260 when they were studied in 1961, all descend from fifteen individuals who, starting in 1816, settled on the island during the nineteenth and early twentieth centuries (Harris *et al.*, 1963). Since, however, the condition exists in about 5 per cent of an English sample of unrelated men and women, it seems that we have here a polymorphic phase which has merely increased its frequency owing to the relaxed selection-pressure to which it would be subjected in a small but expanding community (Ford, 1971, Chapter 2).

True, or balanced, polymorphism always involves an equilibrium between opposed selective agencies. As already pointed out, a form will usually possess some advantage if it has spread through a population to any considerable extent. But normally this will lead merely to a transient polymorphism. Stability will be attained only when the advantage conferred diminishes, and is finally converted into a disadvantage, as the form becomes proportionately commoner. There are instances in which it is 'ecologically' undesirable that a population should become uniform and that the different phases which comprise it should depart from definite proportions which constitute their optimum. Thus the existence of the two sexes is a situation which falls within the definition of a balanced polymorphism. In each species they are maintained at an optimum ratio which is generally, but not always, near equality. We may be confident that any modification of the sex-determining mechanism tending to increase the frequency of one sex at the expense of the other would be opposed by selection.

The check to the progress of an initially advantageous gene will, on the other hand, usually be genetic. Occasionally it will be imposed automatically by 'frequency dependence' acting in favour of whichever morph becomes the rarer. This will arise, for instance, when birds form a 'searching image' and tend to look for a form of prey similar to the one they last secured. Alternatively, the spread of an initially favoured gene will be stopped at some stage by the evolution of heterozygous advantage ('heterosis'). This can arise in either of two ways. It may do so as the result of dominance-modification in the following manner. Genes probably always have multiple effects and, since it is much easier to damage than to improve a complex system such as an animal (or plant) body, if one of them is advantageous the others will probably be harmful. Selection will tend to make any advantageous effect of a gene dominant and its disadvantageous ones recessive. The heterozygote will then possess only advantageous characters, while the two homozygotes will give rise to some that are advantageous and to others that are disadvantageous; consequently there will be selection for the heterozygote, which will lead to stable polymorphism (Fisher, 1930a, Sheppard, 1953a).

Secondly, heterozygous advantage can accrue owing to linkage. Polymorphism ensures that the gene producing the dominant phase will seldom be homozygous. Now mutations frequently give rise to the recessive lethal or semi-lethal condition, the genes for which can accumulate in the immediate neighbourhood of the dominant 'switch gene', because they will be maintained with it in the heterozygous state and so sheltered from selection. Further along the chromosome they will too often be separated from it by crossing-over to gain such protection and will be eliminated when homozygous (Ford, 1945, Sheppard, 1953b). Thus the advantage of the heterozygote compared with the homozygous dominant may also arise in polymorphism owing to the recessive lethals which the switch gene tends to accumulate in the section of chromatin close to it.

In recent years, gel electrophoresis has shown, specifically and by inference, that organisms in general, including man (Harris, 1966), possess a large amount of protein variation much of which is polymorphic. Lewontin and Hubby (1966) hold that heterozygous advantage (heterosis), which involves the relative elimination of homozygotes, may be an effective mechanism for maintaining polymorphisms when they are few but not when they are numerous. For the amount of 'genetic death' involved would then be too great for the organism to bear without undue decline in fitness. They therefore suggest, in spite of the conclusions to the contrary discussed on pp. 111–12, that protein polymorphism is maintained by mutation and by genes of neutral survival value. However, Milkman (1967) points out that this view is wrong because it is based upon two fundamental errors. First, that the genes controlling distinct polymorphisms operate independently, so that their fitness can be combined as a product; which is inadmissible in view of genetic interaction. Secondly, that the unit of selection is the gene, whereas it is the individual. In fact, the result of Milkman's calculations shows that, when these two mistakes are corrected, heterosis is acceptable as a major cause of heterozygosity in nature.

A much quoted paper by Haldane (1957) also makes the mistake of multiplying the effects of the genes as a product. Therefore the evolutionary dilemma which he envisages in it is illusary.

The view that protein polymorphism is selectively neutral and maintained by a high mutation-rate and random drift has been put forward by Kimura (1968) and his associates (e.g., Kimura and Ohta, 1971). Much direct evidence is accumulating to show that this concept is fallacious (Ford, 1971). Thus Powell (1971) studied twenty-two polymorphic loci controlling protein variation in *Drosophila willistoni*. He found that the average heterozygosity and number of alleles per locus was higher in populations kept in heterogeneous than in constant environments. Thus some at least of such polymorphism is maintained in response to environmental diversity and consequently is not neutral as to selection. In attempting to support the contrary view, it has even been suggested that some of the selective effects disclosed by Powell's investigations could apply to the inversions within which the controlling genes may be situated rather than to the loci themselves. But this is unreasonable, for the selective advantage of an inversion is a property of the genes within it.

This and other studies which should have demonstrated the selective neutrality of protein polymorphism, if that be a reality, have notably failed to do so: even, as shown in the striking analysis of Bulmer (1971), in a character specially selected by Kimura and Ohta themselves in order to provide a clear illustration of their views. One may instance also the demonstration given by Bullini and Coluzzi (1971), working on the mosquito *Aedes aegypti*. They examined a locus (*Pgm*) with a large number of electrophoretically detectable alleles. Although the stocks which they used were of most diverse origins (from Africa, Asia, America and the Pacific

Islands) the *PgmA1* allele proved always to be of the highest frequency in 17 out of 19 populations and the only one present in 3 non-polymorphic samples. These results are quite inconsistent with selective neutrality of the polymorphism.

It is of course possible that instances cited for or against that concept may be special cases rather than illustrations of a situation generally applicable. That aspect of the matter has been examined by Harris (1971) in a paper reviewing the control of protein polymorphism in general. He attempts, seemingly with success, to be impartial. Yet a study of his analysis shows that the view of Kimura and his colleagues is at the best subject to great difficulties. It is indeed being widely disproved.

It is a fundamental property of genetics that, assuming equal viability and random mating, the three genotypes of a pair of autosomal alleles are distributed in a mixed population (unless it be very small) as $p^2 : 2pq : q^2$, an equilibrium reached in a single generation. That relationship is sometimes referred to as the 'Hardy–Weinberg law', though the great mathematician G. H. Hardy (1877–1947) was not particularly flattered to find his name attached to a statement so mathematically trivial. He did not, however, realize that, if mating be at random relative to the alleles concerned, then departure from these relative frequencies demonstrates that differential selection is acting upon the characters they control. Thus if 5 per cent of a population be recessives, it is important to know if the heterozygotes differ significantly from their expected value (34·7 per cent of the total).

Making similar assumptions in regard to viability and random mating, the gene-frequencies with total sex-linkage are $p : q$ for the heterogametic sex and $p^2 : 2pq : q^2$ for the homogametic one (see p. 53), though here equilibrium is not reached in a single generation. In these circumstances, the frequencies of a totally sex-linked recessive in men and women will be as $q : q^2$. Therefore if one man in 12 is affected by deuteranomalous green colour-blindness (p. 124), one woman in 144 should be so.

5.2 POLYMORPHISM IN MAN

The existence of a polymorphism is a challenge to the geneticist and, in the human field, should be so to the clinician. For it advertises the presence of something important to the whole population in which it occurs: that an advantageous quality is held in equilibrium by some disadvantage when the polymorphism is of the balanced type, or is spreading unchecked when transient; though the latter situation has not yet been studied in man.

The human polymorphisms may conveniently be introduced by examples which reveal their essential qualities.

The haemoglobins contain two peptide chains. Both are paired and each of the four so formed is associated with a haem group. It is the peptide chains which differ in the various types of this respiratory pigment.

The normal adult form HbA, consists of a major part A1 and a minor, A2. During late embryonic life it has begun to replace foetal haemoglobin, HbF, which still predominates at birth but has almost disappeared four months later. The HbA1 molecule is composed of two α and two β chains; that of HbA2 of 2 α and 2 δ ones, while HbF is the product of 2 α and 2 γ chains. A single locus controls each pair, one gene being responsible for the two identical units. It will be noticed that a mutation of the gene for the α chains will affect all three haemoglobin types since that particular globin occurs in each of them. On the other hand, a mutation of the genes responsible for the β, δ or γ chains can affect one type only.

Weatherall (1971) makes an important generalization in regard to inherited diseases affecting haemoglobin production, pointing out that these fall into two main classes: (1) structural abnormalities of a peptide chain, sickle-celled anaemia belongs here; (2) errors in the rate of formation of one or more chains, from which arises thalassaemia. These two types must be considered separately.

Sickle-cell anaemia is due to a mutant gene at the locus responsible for the β chain. This results in the production of sickle-cell haemoglobin (HbS), in place of the HbA1 type. The mutant acts by the substitution of valine for glutamic acid at a fixed position on the chain.

In those who are homozygous for the sickle-cell gene, haemoglobin A1 is wholly replaced by the sickle-cell form (HbS/HbS). This modifies the structure of the erythrocytes, for they assume a sickle-like shape with long filamentous processes; it also leads to haemolysis severe enough to cause fatal anaemia. Incidentally, such individuals also possess a small amount of foetal haemoglobin, which should have disappeared in the first year of life.

In the heterozygotes, on the other hand, about half the haemoglobin A1 is of type S; but the remainder is of the ordinary form, the condition being HbA1/HbS, and this is sufficient to ensure that the shape of the erythrocytes should be normal when in circulation. However, they assume the sickle-like appearance, though to a less pronounced degree, if the oxygen-tension of a drop of blood be lowered by excluding air from it or by the use of ascorbic acid or other agencies. In this way, the heterozygotes can be distinguished, and they are found to be perfectly healthy. It will be noticed that the anaemia is recessive while the formation of the abnormal haemoglobin is intermediate in the heterozygotes. Moreover, as will be explained, the condition gives resistance to a form of malaria, and the protection so gained is dominant (see also pp. 119–20): a demonstration of the fact that dominance and recessiveness are properties of characters not of genes.

A benign form of sickle-cell anaemia has been detected in two oases in Saudi Arabia. This has lately been intensively studied in a group of 18 Shiite Saudi Arabians carrying the sickle-cell gene (Perrine et al., 1972). There is no doubt that all are homozygotes since their HbA1 is wholly absent. They have foetal and sickle-cell haemoglobin, the latter identical with the African, and the normal amount of HbA2. It has been held that in the homozygotes the symptoms are incompatible with reaching adult life; and generally this is so in Africa, where they are very severe. Yet the present sample of Saudi Arabians ranges in age from 11 to 56, five being 40 or over while eight are 35 or over. Three of the women have had a total of fifteen pregnancies going to term without complications.

None was incapacitated by the disease; indeed the only symptoms which seemed relevant to it were mild attacks of musculo-skeletal pain in about half the group, while a few had an enlarged spleen in their earlier years: indications that HbF is at some disadvantage after birth compared with HbAi, even when selectively adjusted.

It is clear that these homozygotes are viable in the total absence of HbAi owing to their ability to form large amounts of HbF, which ranged from 4·0 to 33·0 per cent (the mean = 18·9). In the normal, severe, condition of the homozygotes, as found in Africa, some trace of foetal haemoglobin persists. Doubtless the amount is genetically variable, and evidently it is upon this that selection has worked to produce the mild condition of the homozygotes that exists in Saudi Arabia: a remarkable example of the way in which organisms can at need sometimes repair their defects by exceptional means, evolution being opportunist.

In spite of the fact that the homozygotes suffer from a heavily lethal disease which usually eliminates them, the heterozygotes are quite common in certain regions. They comprise 17 per cent of the population in some parts of Greece, and as much as 40 per cent in some African tribes. Evidently the heterozygotes must possess an advantage which strongly counter-balances the destruction of the homozygotes in these districts. The nature of this advantage has been discovered (Allison, 1954), for the sickle-cell trait confers marked immunity against malaria due to *Plasmodium falciparum*. The polymorphism is established only in those places where the malaria is common: thus in Europe it occurs in parts of Greece and Italy. Allison, working in the neighbourhood of Kampala in East Africa, found individuals with the sickle-cell trait relatively free from infection. Using native children between five months and five years of age (so as to avoid the difficulties introduced by strong acquired immunity), 113 out of 247 (45·7 per cent) with normal blood were infected, but only 12 out of 43 (27·9 per cent) with the sickle-cell trait were so. The difference is heavily significant. Allison confirmed this result by a method rare in human genetics, that of direct experiment. He artificially inoculated with *P. falciparum*, 15 normal and 15 sickle-celled natives. These were adults, and 14 of the normal group contracted the disease while only two of those with sickle-cells did so.

The association of this blood condition with malarial regions, and the extreme frequency in the population there of a gene nearly lethal when homozygous, is now explained: the abnormal erythrocytes of those with the sickle-cell trait are less easily parasitized by the *Plasmodium*. Thus where malaria is common we have a polymorphism in which the advantage of a gene which confers increased resistance to the parasite is balanced against its lethality when homozygous (see also p. 118). Though the evidence is not very clear, there seems general agreement that while the sickling trait protects against *Plasmodium falciparum* it does not do so for other species of the parasite.

The polymorphism of haemoglobin C, found commonly in the native populations from the Gold Coast to Sierra Leone, is related to the sickle-cell type. It is controlled by yet another allele of the gene for HbA1 and results from the replacement of the glutonic acid at the same site as in HbS, but by lysine instead of valine. The heterozygotes HbA1/HbC are also at an advantage, though this does not seem to consist in protection from malaria, but perhaps from some other blood parasite. On the other hand, heterozygotes of the type HbS/HbC are heavily handicapped since they possess no normal haemoglobin A1.

We may now turn to the thalassaemias, as exhibiting the other class of disease involving haemoglobin production (p. 117): that due to an error in the rate of forming peptide chains rather than in their structure. There are two chief thalassaemia types involving, respectively, the α and β chains. When a mutant reduces or annuls α chain synthesis it of course affects all three normal haemoglobins: HbA1, HbA2 and HbF. The mutant delaying β chain formation naturally involves HbA1 only.

In both α and β chain thalassaemia, the individual may be heterozygous, when the effects are not serious ('thalassaemia minor') or homozygous, giving rise to the usually fatal 'thalassaemia major'. In the latter condition, there is severe anaemia associated with an abnormality of the erythrocytes: they are thin, almost annulet-like in appearance but not sickle-shaped. There is also severe leucocytosis which, together with the heavy erythrocyte elimination, leads to over-development of bone marrow with consequent reduction in the cortex of the long bones.

Here again the condition is thought to be inimical to the development of malaria parasites, which gives an advantage to the thalassaemia minor cases in malarial regions and is responsible for the polymorphism of the disease there. There is indeed a positive correlation between the distributions of malaria and thalassaemia. In some parts of Greece and Italy, the gene for β thalassaemia, the one generally found there, may be present in 10 per cent of the population. The form principally established in the Near and Far East is α thalassaemia, in which Weatherall (1971) shows that there is a total deficiency of α chain synthesis in the homozygotes. Indeed his brilliant analysis of the molecular basis of the thalassaemias demands careful study.

It is appropriate at this point to mention the polymorphism of the haptoglobulins, which are proteins concerned with the binding of haemoglobin released from broken down erythrocytes. Two alleles, recognized by electrophoresis, control their variation. Hp^1Hp^1 produces a single haptoglobulin type while Hp^2Hp^2 gives rise to a considerable series. Moreover, a further set, differing from those evoked by either homozygote, arises from the heterozygotes Hp^1Hp^2: an instance of the fact that the genes may interact to produce the features for which they are responsible. Since the genotypes are distinguishable, while in England their frequencies are approximately Hp^1Hp^1 17 per cent, Hp^1Hp^2 48 per cent, Hp^2Hp^2 35 per cent, they should be useful in paternity tests.

The taste-test provides another aspect of human polymorphism, while our knowledge of the contending advantages and disadvantages which maintain it is more a matter of inference. The arguments upon which this is based are worth consideration since some of them are of a type to which recourse must often be made in the study of human genetics; however, the nature of the experimental proof involved was unique when this book was written in 1942 and has indeed remained so.

The ability to taste phenyl-thio-urea in low concentrations[1] has already been discussed as a simple example of the working of Mendel's first law (pp. 3–11). It will be recalled that those who can detect this substance possess a gene T dominant in effect, while the inability to do so is recessive and due to the operation of an allelomorphic pair tt. When both parents are non-tasters, so are all their children. When one parent is a taster and the other is not, either all or half their children will be tasters too: depending on whether or not the parent of the dominant (tasting) type is a homozygote or a heterozygote. Similarly, when both parents are tasters, so will be all, or else three quarters, of their offspring. The applications of these facts will be discussed further on pp. 166–73. Non-tasters occupy from 20 to 30 per cent of the population in Europe and the Near East, the proportion varying within these limits according to race (Boyd and Boyd, 1937). It seems, however, that they are much less common among Chinese (about 10 per cent) and African Negroes (about 3 per cent), though the samples on which these estimates depend are small (Barnicot, 1950).

Numerous human sensory and other defects are known, or presumed, to be inherited in various ways, and the inability to taste phenyl-thio-urea is among the most interesting of them, on account of its extraordinary frequency. For this is a unifactorial condition in which the two allelomorphs concerned seem to be in approximate equality at least throughout Europe and in populations of European origin. We cannot here be dealing with a disadvantageous gene subject to elimination and maintained by the rare process of recurrent mutation. On the contrary, all the circumstances indicate that this situation is an example of balanced polymorphism. Alternatively, it might be suggested that it is immaterial as regards survival value which of the two types of allelomorph is operating. Without due consideration, this latter view might unhesitatingly be accepted. It must be supremely unimportant whether or not a man can taste phenyl-thio-urea. Indeed no one had the opportunity of doing so until the present century. It

---

[1] 'Tasters' can detect it in aqueous solutions of 50 parts per million while 'non-tasters' cannot. Both types can usually do so in concentrations of 400 parts per million. There are other slight complications.

might well be held that the frequency of the condition observed in man is due merely to the chance spread of a gene controlling a character of no survival value. On the other hand, it has been stressed that genetic factors have multiple effects, and it may be expected that some of those produced by the gene responsible for this taste-distinction will probably be of significance for the welfare of the individual. The theoretical considerations so far outlined make it certain that this is so.

Fortunately, however, we are no longer dependent upon theory alone in order to reach this conclusion. For the ability to taste phenyl-thio-urea provided the first evidence for the reality of such polymorphism in man. This is of sufficient importance to merit a brief description.

It appeared just conceivable that a study of the anthropoid apes might provide decisive evidence that a long-established polymorphism is at work in maintaining at optimum proportions the genes controlling this taste defect. The possibility of such an approach was realized, and tested successfully, by Fisher, Ford and Huxley (1939). They found that 'tasters' and 'non-tasters' exist in at least two species of ape, the orang-utan and the chimpanzee. These animals can be classed for this character without risk of confusion, and their level of perception proved to be approximately the same as that of man. Using the chimpanzee, it was found that out of 27 individuals examined, 7, or 26 per cent, were non-tasters, which strongly suggests that the condition is a balanced polymorphism in them also. It is indeed hardly conceivable that the special mechanism controlling such an unexpected character should have arisen by chance independently in the anthropoid and hominid stocks. On the contrary, we here have evidence that the advantage of the heterozygote in the common ancestor of man and the chimpanzee has been preserved in the evolving lineage of both up to the present day. Wherein the advantage lies it would at present be idle to speculate, except that it is presumably of a physiological nature. However, of its existence there can be little doubt.

The above passage, predicting that the taste-test genes are not of neutral survival-value, was written in 1941 and published in

GENETICS FOR MEDICAL STUDENTS

1942 in the first edition of this book. Evidence demonstrating its truth is now available. In 1949[1] it was shown that there is an association between this polymorphism and liability to thyroid disease. For a review of this subject, see Clark, 1964, pp. 217–19.

In addition to the blood group systems and related conditions to be described in the next three chapters, it will be realized from their frequencies that some of the unifactorial or super-genic characters already mentioned must be polymorphic. A large number of human polymorphisms have now been recognized: a few only of these need be mentioned here. They are discussed by Clarke (1964) and Ford (1965, 1971), while several of them are listed in Appendix II of this book. Reference may, however, be made to the rapid and slow inactivation of the isoniazid drugs, the two types being in approximate equality in two Caucasian and two Negro populations studied by Evans et al. (1960); to the blue- or grey-eyed types compared with the brown-eyed; and to the inability to smell freesia flowers, which seems to affect about 8 per cent of the community in England. Furthermore, two of the other polymorphic systems found in Man must be considered in rather more detail at this point.

In the first place, colour-blindness, though discussed on pp. 53–5, merits additional study, from this point of view. As already explained, it is approximately recessive and behaves as if determined by a single gene, totally sex-linked to X.

However, there are in fact two loci involved, controlling respectively ability to see red (protanoids) or to see green (deuteranoids). These types can be subdivided further. There are protanoids who cannot see red but can see green and blue (protanopia), and those who can see all three but require an excess of red light to combine with the other two so as to give the effect of white (protanomaly). They are determined by alleles at the protan locus. The deuteranoids fall into two similar subdivisions, also allelic: so that they include deuteranopia and deuteranomaly.

The latter is the commonest of the four, and accounts for about 60 per cent of all those who are colour-blind. As Clarke (1969)

[1] Harris, H., Kalmus, H. and Trotter, W. R.: (1949), *Lancet*, 2, 1038.

points out, the term 'red-green colour-blindness', generally applied to all these defects, is inaccurate since the individuals concerned always suffer from red or green blindness, not both. There seems some evidence for thinking that while the loci determining the two conditions are fairly near one another (with a C.O.V. of about 9), they are separated by that for G6PD deficiency (pp. 108–9), so that they do not form a super-gene.

Post (1962) reviews the anthropological aspect of colour-blindness. He finds that the defect is rarest (0·02 of males) in primitive peoples such as the aborigines of Australia (0·019), Brazil, Fiji and North America, and commonest (0·08 of males) in civilized communities: Europeans, Brahmins in India and the cultured inhabitants of the Far East. He suggests that this is a consequence of relaxed selection against the condition, which may well have taken place since the beginning of the Neolithic Period, perhaps 120 generations or so. For this is a failure in perception which might certainly be disadvantageous among hunters and food gatherers.

That conjecture seems very reasonable and is probably correct. However, Post then goes on to suggest that this difference in frequency between the two types of community, wherever they respectively occur, is due solely to mutation. For he takes the view that colour-blindness is of neutral survival-value in the civilized groups, so that its spread is no longer kept in check by counter-selection; though he himself points out that the affected individuals can detect certain visual differences indistinguishable to normal people: a fact employed in the Ishihara tests. He even uses the excess of affected persons in civilized conditions to calculate the mutation-rates at the protan and deuteran loci; an attempt which forces him into the astonishing position of 'discounting reverse mutations'! If the alternative alleles were really of neutral survival-value save in hunting and food-gathering societies, one might indeed wonder why the frequency of colour-blindness had always *increased* in higher cultures: it should have wandered haphazard to greater and lesser levels, unless evidence be produced that mutations towards the one allele are substantially commoner than towards the other. Indeed Post himself seems uncertain of his position, since he remarks 'if the genes for colour-blindness are supposed to have an adaptive advantage in the new habitat, the resulting evolutionary change would probably be attributed to progressive adaptation not to relaxation (i.e. of selection)'.

It should be stressed that colour-blindness must have advantages and disadvantages in both types of environment. We are not in this matter merely postulating unknown pleiotropic effects. The polymorphism concept itself demonstrates their existence, for the genes could not have attained their present frequencies without them. A relaxation of selection has merely shifted a balance of advantage and disadvantage to a new and higher value.

Another important aspect of genetic variation in man, one which includes polymorphism, is provided by intra-ocular pressure. This has been the object of intensive study in recent years. It has been brilliantly surveyed by Evans (1971), who provides detailed references to each aspect of the work.

Variation in intra-ocular pressure in healthy eyes is unimodal, falling within a curve of normal distribution, a study of which demonstrates that it is under polygenic control. With the use of corticosteroid eye drops, however, it becomes polymorphic. It then falls within three groups, the two homozygotes and the heterozygotes, resulting from the segregation of a pair of alleles P^L and P^H. Their ratios are approximately $1:2:1$ among the offspring of the cross $P^L P^H \times P^L P^H$; while the frequencies of the phenotypes are $P^L P^L$ 66, $P^L P^H$ 29 and $P^H P^H$ 5 in the general population of the U.S.A. It has been shown that these two alleles contribute to the ordinary polygenic variation mentioned above, but that they exert an over-riding influence in the environment produced by corticosteroid drops, when they generate a polymorphism.

Moreover, these genotypes differ in their liability to induce open-angle glaucoma (raised intra-ocular pressure in the absence of of any causative structural defect) in later life. Thus Armaly (1967a) calculates that if the risk be $1 \cdot 0$ in a $P^L P^L$ individual, it is $18 \cdot 0$ for the $P^L P^H$ type, and $101 \cdot 0$ for $P^H P^H$.

There is thus a hereditary component in the liability to this disease; though a recent twin-study suggests that it may have been over-stressed (Schwartz *et al.*, 1972), especially in the environment produced by corticosteroid eye-drops. However, all glaucoma patients should certainly be instructed to warn their children, especially their sons since the condition is to some extent sex-controlled to the male, to have the pressure in their eyes tested three times a year from, say, the age of forty onwards. I am well acquainted with an instance in which this resulted in the detection of that disease at an exceptionally early stage when it could be effectively controlled by medication (e.g. pilocarpine drops) before extensive retinal damage had been done. For it must be noticed that the advance of glaucoma is most insidious, often reaching a dangerous state, when much irreparable destruction has taken place, before attracting the patient's attention.

There is an association between diabetes mellitus and glaucoma, the frequency of the P^H allele being greater in diabetic patients (Becker *et al.*, 1966). This might be thought a spurious effect for the following reason. Owing to its insidious nature, there are large numbers of unrecognized glaucoma cases in the general population. They will only be detected in the more terminal stages which, moreover, a proportion of these people will never reach owing to death. It is probable that diabetics may be subject to more thorough general medical examination than those who are healthy, resulting in the discovery of unrecognized glaucoma in them. Once the suggestion of an association between the two conditions has been raised, the intra-ocular pressure of diabetics will generally be tested; a process which would increase the apparent frequency of glaucoma among them even if it were in fact no higher than normal.

Though this must surely contribute to the observed correlation between glaucoma and diabetes, it provides only a partial explanation of it. Armaly (1967b) studied a group of 55 individuals with clinically normal eyes (mean age, 25 years) who had their P^L and P^H genotypes determined by a dexamethasone test. At another time, their oral glucose tolerance was assessed, and this showed that the P^H gene provides less glucose tolerance than P^L. We have here a genuine basis for an association between glaucoma and diabetes: one which does not depend upon detecting normally unrecognized cases of the eye disease among the diabetic patients.

Instances are known of multifactorial inheritance with one of the genes involved having a disproportionately great effect; sometimes in certain environments only, genetic or physiological (as, apparently, in glaucoma). Evidence exists for the genetic control of essential hypertension: some studies suggesting that it is a multifactorial condition, others that it is unifactorial and 'dominant'. These contrary conclusions can perhaps be resolved along the lines just indicated.

A final proposition must be stressed here. That is to say, all polymorphisms, even if apparently trivial, are maintained by a balance of selective advantages and disadvantages. They must therefore have additional effects, often cryptic in nature, that are of importance to the individual.

CHAPTER 6

The Blood Group Polymorphisms:
The Simpler Systems

6.1 INTRODUCTION

The erythrocytes carry substances known as *antigens* each of which can react with a corresponding *antibody* present in the plasma. When they do so, the cells of the blood clump together, or agglutinate, and undergo haemolysis. It is obvious therefore that a corresponding antigen and antibody cannot normally coexist in the same individual. The antibodies may be *spontaneous*; that is to say, they occur naturally, being manufactured without extraneous stimuli. Alternatively, they may arise as an immunity reaction, when they are described as *immune*. They are then produced in response to the presence of an inappropriate antigen. This may reach the blood by way of the placenta in women or in a transfusion in either sex.

Consequently, a certain risk is associated with multiple transfusions, even from the same individual whose blood had proved satisfactory when first used. For it might have stimulated the production of an immune antibody which, persisting in the circulation, could react disastrously with its corresponding antigen on the second occasion. This condition has been referred to as 'anaphylaxis' in medical textbooks. Transfusion after a pregnancy when, of course, such treatment may be particularly needed, is subject to a similar danger, owing to possible antibody formation in the maternal blood in response to antigens present in that of the foetus. The transfusion reaction so produced may give rise to a series of complex symptoms. When incompatible erythrocytes are introduced, they may be haemolyzed within the blood vessels, as with anti-G(A) and anti-G(B), p. 147. The other blood group antibodies are, however, removed intact by the tissues of the reticulo-endothelial system chiefly within the liver and the spleen.

GENETICS FOR MEDICAL STUDENTS

It does not seem that intravascular agglutination choking the blood vessels can be very important, although formerly thought to be so. The most characteristic symptom of incompatibility is constricting pain in the chest and back. This is probably due to the release of active substances into the circulation.

Blood group antibodies may exist also in a second form, one in which they do not usually agglutinate their corresponding antigens when erythrocytes carrying them are suspended in saline. These are the so-called 'incomplete antibodies'. Yet we know for several reasons that they are capable of serological reaction. First, the incomplete antibodies do agglutinate the appropriate types of red cells when suspended in certain protein media. Secondly, they will do so even in saline, as the normal antibodies always do, if the red cells have been incubated with certain enzymes. Thirdly, it can be shown that red cells in saline are in fact sensitized to the appropriate incomplete antibodies, which can be done by the addition of anti-sera made by rabbits against human globulin.

Individuals who are alike in possessing or lacking a given antigen, or antigens, belong to the same blood group. These *groups* fall within a number of distinct *systems*, the antigens and antibodies of which do not cause agglutination with one another. The occurrence of the various groups of a system is controlled genetically as a balanced polymorphism: that is to say, by genes or super-genes which act as a switch in determining the alternative types, and these must be maintained in the population by counter-balancing advantages and disadvantages. Even the rarer of the allelomorphs concerned are therefore far above the frequency which can be sustained by mutation. There are, however, some instances in which an antigen is not polymorphic at all but is constantly eliminated by selection. The frequency of the gene concerned is therefore so low as to approach its mutation rate (pp. 67–8). This is merely an example of rare genetic variation, like albinism.

The naturally occurring antibodies of the ABO system exist in all individuals in whom they are compatible. The blood groups involved are therefore of fundamental importance in transfusion. So too are those of the Rhesus system, because certain of its antibodies are very easily induced as immune reactions, although they are rarely found spontaneously. It happens, however, that the complications both of ABO and Rhesus are considerable. Consequently, the genetics, and indeed the serology, of the blood groups are more easily understood by considering first some of the simpler types, although they do not often give rise to haemolytic disease. These will, therefore, be described in this chapter. They will form an introduction to the systems of greater clinical importance which are reserved for treatment in Chapter 7, when they will be more easily followed.

For much of the data in this and the next chapter, I am indebted to Race and Sanger (1968), who have provided a masterly survey of the human blood groups. Their own brilliant researches, and those of their colleagues at the Lister Institute, London, have contributed to an outstanding degree to our knowledge of serology. For information on the distribution of the blood groups in the various races of mankind, I have consulted the remarkably detailed book on this subject by Mourant (1954).

6.2 NOTATION

The chief barrier of the study of the human blood groups has been their chaotic notation. It has increased immensely the difficulty of explaining and understanding both their genetics and their serology. Not only has it been out of accord with current genetic usage but, far worse, it has been inconsistent from one series to another and is actually misleading. Moreover, surprising as it may seem, even in the most authoritative textbooks on the subject, there is frequent uncertainty as to whether a given symbol refers to a gene or an antigen; while the complex and unusual dominance relationships characteristic of the blood groups have been obscured and falsified by the methods of representation.

I have, therefore, introduced a simple and uniform notation for the human blood groups (Ford, 1955).[1] While I should be the last to claim perfection for it, this will, I hope, mitigate the

---

[1] The editors of *Heredity* have been so good as to allow me to use, with little alteration of wording, some passages from the original account of this notation.

present confusion and render the subject easier to understand and to remember. It is, of course, adopted in the account which follows. I propose at the outset to explain it in general terms; its subsequent application should then present little difficulty.

(i) GENES

These are always to be represented in italics. Symbols for antigens and antibodies should never be italicized.

As elsewhere in genetics, we are concerned with various loci, at each of which different gene-substitutions can take place. Ordinarily, one of the allelomorphs is dominant and another recessive; that is, in respect of the character by which we identify them. As already explained, it is the custom to represent the dominant by the capital and the recessive by the small form of the same letter; for instance, the gene for recessive albinism (*a*) and its dominant allelomorph for normal colouring (*A*). The notation used for multiple allelomorphs has been explained on pp. 23-4. In serology, however, it is usual to find that both members of an allelomorphic pair produce distinct antigens, each of which may be fully dominant (having a similar effect in single and double dose), or, at least, having detectable effects in the heterozygote as well as in the homozygote. Clearly such a notation as *A*, *a* is inappropriate to that situation though, surprisingly, it has been so used by serologists in several blood group systems. Moreover, when an allelomorph is not known to produce an antigen, it is more probable that no antibody detecting it (by agglutination) has yet been found than that it does not exist. The notation, therefore, must be one which can be expanded to accommodate new discoveries. To meet this somewhat exceptional situation, the following plan has been devised.

The *locus* of a gene is invariably indicated by a capital letter, or by a pair of letters of which the first is a capital (examples: *K*, *Lu*). The *allelomorph* is indicated by the presence or absence of a suffix attached to the locus-symbol. When allelomorphs, each represented with capitals in the suffix, are brought together, they both exercise their effects (one dose of a gene may or may not be as effective as two). A capital in the suffix is dominant to a small letter in the suffix. Examples: K^AK^A, the antigen produced by the gene K^A is present on the erythrocytes; K^BK^B, that produced by K^B is present; K^AK^B, both antigens are formed.

A gene without a suffix is one whose antigen has not yet been recognized by the discovery of its corresponding antibody (or conceivably it is not formed). Such a gene is recessive to one represented with a capital in the suffix, but dominant to one with a small letter in the suffix. Examples: Do^ADo^A and Do^ADo: both genotypes give rise to the antigen produced by the gene Do^A. No antigen formed by the homozygote $DoDo$ has yet been detected.

When the members of a multiple allelomorph series are distinguished by numbers, a large figure in the suffix represents a gene dominant in effect to one with a small figure. Example: G^{A1}, G^{A2}, G. The first of these is dominant to the second, and both are dominant to G, as the foregoing notation indicates. A natural extension of this arrangement covers other instances in which the effect of one gene is dominant to another while both are dominant to a third, as will be seen later in dealing with D^U of the Rhesus system.

When two or more genes are so closely linked as effectively to act as a single switch-mechanism, a condition which tends to be evolved in polymorphisms such as the blood groups, they are to be represented not only on one side of a line, as with ordinary linkage, but they are, in addition, to be placed together within a bracket.

(ii) ANTIGENS

The letter or letters of the locus-symbol and the suffix letter used for the allelomorph, but unitalicized, become transformed into the discriminators for the antigens and antibodies. The way in which this is done prevents any confusion between genes, antigens and antibodies.

The antigen group is shown by the locus-symbol. The suffix-letter follows within brackets and shows the types of antigen present; those absent are also indicated when necessary. When the *presence* of an antigen is dominant, a capital is used for the corresponding letter within the bracket; when its presence is recessive, a small letter is used. The presence of an antigen is shown by a plus sign following the discriminating letter within the bracket, its absence by a minus.

Examples: The Kell antigens are: K(A+), K(A−); or, if known, K(A+B−), K(A+B+), K(A−B+). Both these antigens are dominants.

This antigen notation refers to the presence or absence of given antigens on the red cells. It may also be necessary to refer to the antigen itself, such as the presence of an antigen in the saliva (pp. 152–3). In such circumstances, the plus or minus sign is not relevant and can be omitted. Thus it is possible to speak of the K(A) antigen.

(iii) ANTIBODIES

These are indicated by the locus-symbol with the prefix 'anti' hyphened to it, followed by the gene-suffix in brackets. Example: anti-K(A).

6.3 THE SIMPLER BLOOD GROUPS

Though the blood groups included in this chapter are of relatively little clinical importance, owing to the rarity of their antibodies, whether spontaneous or immune, they are of much interest in other fields: those of genetics and ethnology. Moreover, since they are polymorphic, they must, as already explained, be maintained by a balance of selective advantages, though the nature of these is not fully known. Since even the rarer of the genes involved in controlling them must occupy a considerable proportion of the available loci, they can act as useful markers of the human chromosomes (pp. 182–3). The various series in this and the succeeding chapter are arranged in increasing order of complexity, in the light of present knowledge. This should make it easy to understand them.

(i) THE DUFFY SYSTEM

GENES: Fy^A, Fy^B, Fy.

ANTIGENS: Fy(A+), Fy(A−); or, if known, Fy(A+B−), Fy(A+B+), Fy(A−B+), Fy(A−B−).

ANTIBODIES: anti-Fy(A), anti-Fy(B).

Approximately 65·7 per cent of the population in England and Wales carry the antigen Fy(A) including those homozygous and heterozygous for the gene controlling it, and these amount to 17·2 and 48·5 per cent respectively. These genotypes can on the average be distinguished by a slight dosage effect, two of the Fy^A genes react with the antibody much more strongly than one. Anti-Fy(A) sera, though rare, have now been found on many occasions, and have repeatedly been responsible for serious haemolitic disturbances on transfusion. The antibody is usually, though not always, immune in origin.

Anti-Fy(B) sera are much rarer than anti-Fy(A), having been found on a few occasions only. Since the corresponding antigen Fy(B) is, so far as known, a complete dominant, it occurs in 82·6 per cent of the English population.

The frequency of the Fy(A) antigen seems to vary greatly from one race to another, though in the more striking instances the numbers so far tested are quite small. Thus twenty-four Chinese all proved to be Fy(A+).

Unlike whites, 68 per cent of New York Negroes are of the type Fy(A−B−), due to the action of a third allelomorph Fy. It is now known that the genotypes Fy^AFy and Fy^BFy occur also among people of European ancestry (Race and Sanger, 1968).

Indeed as explained by Albrey et al. (1971), there appear to be two Fy alleles in whites. Fy itself, which produces no known antigen, and Fy^X which reacts with certain anti-Fy(B) sera and is therefore distinguishable only quantitatively from Fy^B. Albrey et al. also describe an antibody which reacts with all cells except those of the Fy(A−B−) type. Thus the Duffy system proves to be more complex than originally supposed.

(ii) THE KELL SYSTEM

GENES: K^A, K^B.
ANTIGENS: K(A+), K(A—); or, if known, K(A+B—),
K(A+B+) K(A—B+).
ANTIBODIES: anti-K(A), anti-K(B).

The distribution of the three genotypes in western Europe, which does not seem to vary significantly among the few races so far tested, is K^AK^A, 0·21 per cent; K^AK^B, 8·7 per cent; K^BK^B, 91·1 per cent. Thus only 8·9 per cent of the population are K(A+), although this antigen is practically dominant; yet there is a slight dosage effect which causes somewhat less marked agglutination in heterozygotes than in homozygotes. Anti-K(A) sera, though uncommon, have now been detected in many dozens of instances. They have, perhaps, always been immune, and have given rise to marked haemolytic reactions on transfusion. Furthermore, they have several times done so in a foetus whose erythrocytes carry the K(A) antigen. This had reacted with anti-K(A) which had reached the foetal circulation by way of the placenta.

Having regard to the low frequency of the genotype K^AK^A, anti-K(B) sera are, naturally, rare. They identify the K(B) antigen which, being dominant, is carried by 99·8 per cent of the English population. In an extensive test, only 5 out of 2,500 gave a negative response to anti-K(B), which closely approximates to expectation.

There is some evidence of racial differences in the frequencies of the Kell antigen. Information on the subject is, however, at present meagre.

(iii) THE KIDD SYSTEM

GENES: Jk^A, Jk^B.
ANTIGENS: Jk(A+), Jk(A—); or, if known, Jk(A+B—),
Jk(A+B+), Jk(A—B+).
ANTIBODIES: anti-Jk(A), anti-Jk(B).

The frequencies of the three genotypes in England are Jk^AJk^A, 24·8 per cent; Jk^AJk^B, 50 per cent; Jk^BJk^B, 25·2 per cent. Both the antigens are approximately complete dominants, so that 74·8 per cent of the population are of the antigen-group Jk(A+). However, the two antibodies each give slightly stronger reactions against the homozygous than the heterozygous condition. Though the system was only discovered in 1951, a considerable number of anti-Jk(A) sera have now been found, most of them immune. They have given rise to dangerous reactions.

An anti-Jk(B) serum has so far been detected on a few occasions only. It has been responsible for haemolytic disease. One Jk(A—B—) person has been discovered; presumably, therefore, there is also a *Jk* allele.

Ikin and Mourant (1952) report that 95 per cent of West Africans were Jk(A+) out of a total of 105 tested. This is significantly higher than the 76·4 per cent found in the English population. It is probable therefore that this system will prove to be of value in anthropological studies.

(*iv*) THE LUTHERAN SYSTEM

GENES: Lu^A, Lu^B.
ANTIGENS: Lu(A+B—), Lu(A+B+), Lu(A—B+).
ANTIBODIES: anti-Lu(A), anti-Lu(B).

Only 7·7 per cent of the English population are Lu(A+). Since antigen production is dominant, the three genotypes must therefore be distributed as $Lu^A Lu^A$, 0·15 per cent; $Lu^A Lu^B$, 7·5 per cent; $Lu^B Lu^B$, 92·3 per cent. It has been claimed that Lu(A+) is divisible into sharply distinct and genetically controlled sub-groups, as with group A of the ABO series (pp. 149–51). It appears, however, that variations in the Lutheran agglutination fall within a curve of normal distribution and are not maintained as a polymorphism. Less than two dozen examples of anti-Lu(A) sera seem to have been found so far. They have been both spontaneous and immune but they do not appear to have given rise to haemolitic disease.

Since only 0·15 per cent of Europeans are of the type in which anti-Lu(B) can exist (that is to say, $Lu^A Lu^A$) it is not surprising that this antibody is extremely rare. It had not been detected when the fourth edition of this book (1956) was published. Even now only two or three examples of it have been obtained, one of them being responsible for a mild transfusion reaction.

There is evidence that the frequency of the Lu(A+) condition varies in different races, but information on the subject is scanty. It seems clear, however, that in some populations it can occupy a somewhat higher figure (12 out of 73 in Brazilian Indians) than the 7·7 per cent found in Europeans, and that in others it is almost or completely lacking (Indonesians and Australian aborigines).

For linkage between the Lutheran and Secretor genes, see p. 153.

(v) THE Sd SYSTEM

GENES: Sd^A, Sd.
ANTIGENS: Sd(A+), Sd(A−).
ANTIBODY: anti-Sd(A).

This recently described system has caused trouble by the variability of its expression, it being difficult to distinguish between more weakly reacting samples and negatives. New techniques have, however, largely overcome this problem (Race and Sanger, 1968). The antigen is dominant, not recessive as the symbol Sd^a (to be replaced by Sd^A), unfortunately used for it in serology, indicates. Race and Sanger (l.c.) estimate that it occurs in 91·24 per cent of the population in England.

Surprisingly, the antigen is secreted, to a varying extent, in the saliva (pp. 152–3). The antibody, which is not known to have caused transfusion reactions, seems to be found in about 1 per cent of the English population; and it is present in the seeds of the plant *Dolichos biflorus* (p. 152). An antigen described as 'Cad' appears to be identical with this one.

(vi) THE DOMBROCK SYSTEM

GENES: Do^A, Do.
ANTIGENS: Do(A+), Do(A−).
ANTIBODY: anti-(A).

In this system, also recently discovered, the presence of the one known antigen is a dominant character thus, in accord with the ruling of genetics as a whole, its symbol must be Do^A (not Do^a, given to it in serology, since this states that it is a recessive). Tippett (1967) supplies valuable information showing that the Do(A+) phenotype occurs in 66·4 per cent of Northern Europeans. This is approximately the same as the value obtained for Israelis (64·8 per cent) but not for Negroes, in whom it is substantially lower (44·7 per cent). There is no indication of linkage with other known blood groups, and the antibody seems to be a rarity. One may hope, both here and in the Sd System, that another antibody recognizing the alternative allele may be discovered.

(vii) THE P SYSTEM

GENES: P^A, P.
ANTIGENS: P(A+), P(A−).
ANTIBODY: anti-P(A).

The P system requires particularly careful serological work if the groups are to be accurately scored, for quite large variations can arise from differences in the strength of the antibody and in the preservation of the erythrocytes. It seems that P(A+) individuals comprise 78·85 per cent of the population in western Europe and, since the antigen is dominant, the three genotypes are distributed as $P^A P^A$, 29·2 per cent; $P^A P$, 49·7 per cent; $P P$, 21·7 per cent.

The anti-P(A) serum can be obtained from rabbits immunized by human erythrocytes, and in this way it was originally produced. However, it is now known to occur spontaneously in rabbits, as well as in cattle, horses and pigs. In man, a trace of this antibody may be found in nearly all P(A−) individuals. Also, on rare occasions, it is present in a high concentration, having arisen sometimes spontaneously and sometimes as an immune reaction to transfusion. However, there does not seem to be any fully substantiated instance in which it has been responsible for haemolytic disease, probably because it is a 'cold agglutinin'; that is to say, one which is most active at a temperature below that of the body.

It has now been recognized that the P system is much more complex than had previously been supposed though the new dis-

coveries, which can best be thought of in two stages, represent a
refinement which has so far had little practical effect. In 1951 a
new antigen, Tj(A+), was detected. This is found in nearly everyone; indeed those who lack it, and can be represented as Tj(A−),
amount to no more than one in many thousands, and all of them
possess the corresponding antibody, anti-Tj(A). Thus the situation
resembles that of the O, A, B series (pp. 130, 147–53) in
which the antibodies are developed in everyone in whom they are
compatible. The presence of the Tj antigen is inherited as a simple
dominant controlled by the gene Tj^A, its rare allelomorph being Tj.

The next step in elucidating this situation was taken when it
was realized that all Tj(A−) individuals are P(A−). Both this
and other evidence demonstrated indeed that the Tj antigen and
antibody belong to the P system, which may now be expanded as
shown in Table 1.

TABLE I

Groups	Genes	Antigens	Antibodies (compatible)	Antibodies (incompatible)
P1	P^A1	P(A1+A₂+)		anti-P(A1), anti-P(A₂) previously anti-Tj(A)
P²	P^A2	P(A1−A₂+)	anti-P(A1)	anti-P(A₂) previously anti-P(A)
P	P	P(A1−A₂−)	anti-P(A₁), anti-P(A₂)	

The antigen P(A1−A₂−) represents Tj(A−) and the other two
antigens together represent Tj(A+). It will be seen that the whole
situation parallels that of groups A1, A₂, O in the O, A, B series:
see the lower table on p. 151, which was worked out before the
relationship between the P and the Tj groups had been appreciated.

Apparent racial differences in the frequency of the antigens
tend to be unreliable owing to the occurrence of a proportion of
very weak reactors to the P antibody. However, it appears certain
that the frequency of the P(A+) group is significantly high in
Negroes and low in Lapps, Chinese and South American Indians,
compared with Europeans.

(*viii*) THE MNL SYSTEM

(*a*) The M, N Groups

GENES: Ag^M, Ag^N.
ANTIGENS: Ag(M+N−), Ag(M+N+), Ag(M−N+).
ANTIBODIES: anti-Ag(M), anti-Ag(N).

The original locus symbol for this gene was R (Ford, 1942), chosen with reference to the rabbit sera used in the preparation of the antibodies. At that date, the Rhesus system had been discovered less than two years. However, R has subsequently been so much used in connection with Rhesus, that it seems inadvisable to retain it for the locus of the M, N group. Accordingly, Ag has been chosen for this purpose (suggested by Strandskov, 1948).

Ag^M and Ag^N are each dominant, although a single dose of either gene produces slightly less antigen than the corresponding homozygote. Since the antibodies were both easily produced as immune reactions in rabbits, even at the time of the original discovery, the three groups involved have been distinguished from the first and recognized as an example of genetic segregation, in which all possible genotypes can be detected. That the strictly comparable situations, in such systems as Duffy, Kell or Kidd, were not at once understood as such, is due merely to the fact that one of the two anti-sera, necessary for discriminating all three of their genotypes, was, in them, discovered later than the other, and proved rarer.

The three groups controlled by the Ag locus, with their genotypes, and their percentages in western Europe, are:

Groups	Genotypes	Percentages
M	$Ag^M Ag^M$	28·4
MN	$Ag^M Ag^N$	49·6
N	$Ag^N Ag^N$	22·0

Anti-Ag(M) and anti-Ag(N) can, apparently always, be obtained from rabbits which have been injected with human erythrocytes of groups M and N respectively, the cells should also belong to group O of the ABO system (pp. 147–8). The antigens of the other blood group systems do not, in these circumstances, stimulate antibodies, though P(A) may do so. However, this is removed by the necessary absorptions required to get rid of anti-species antibodies that are always formed in addition. The M and N antibodies very rarely occur in human blood, though they have now been detected on a number of occasions, anti-Ag(M) the more often. This seems generally, perhaps always immune in origin, and it has given rise to haemolytic disease on transfusion with blood of groups M and MN. A few instances of anti-Ag(N) in human blood seem to include both immune and spontaneous types. It is not clear that they have been responsible for serious reactions on transfusion. Anti-Ag(N) occurs in the extract of certain plant seeds; for instance, those of *Vicia graminea* (p. 152).

(b) The L Groups

GENES: L^A, L^B.

ANTIGENS: L(A+), LA(−); or, if known, L(A+B−), L(A+B+), L(A−B+).

ANTIBODIES: anti-L(A), anti-L(B).

This locus was formerly represented by *S*, but that letter had been preoccupied for the secretor gene (p. 152) so that it had to be changed. *L* proved to be a suitable substitute for it.

The distribution of the three genotypes in England is L^AL^A, 10·7 per cent; L^AL^B, 44 per cent; L^BL^B, 45·3 per cent. The genes L^A and L^B are both effectively dominants, though a small dosage effect is responsible for slightly stronger agglutination in homozygotes than in heterozygotes. Thus 54·7 per cent of the English population are L(A+).

Anti-L(A) sera, first discovered in 1947, are uncommon, but have now been found on several dozen occasions. Their origin has been sometimes spontaneous and sometimes immune, and it is known that they have produced serious reactions. Only two examples of anti-L(B) sera have so far been found, the first in 1951. Both were immune, and one of them occurred in the blood of a mother whose baby was suffering from haemolytic disease, of which this antibody was the cause.

There is exceedingly close linkage between the *Ag* and *L* loci, which evidently constitute a single super-gene controlling the MNL system as a whole. Crossing-over between them may well be reduced to mutation frequencies. Consequently, the MN and L groups are not distributed at random with respect to one another. That is to say, in England, with 54·7 per cent of the population L(A+), the proportions of this type among the MN groups are: group M, 73·4 per cent; group MN, 54·1 per cent; group N, 32·8 per cent.

Two groups, Hunter and Henshaw, which provide an extension of the MNL series, are almost wholly African (but see p. 171). Our knowledge of them is scanty, but it is clear that both their antigens are dominants. Moreover, the genes controlling them are very closely linked with the Ag and L loci, making an addition to this super-gene. It is not yet certain, however, whether or not they are at separate loci. That is to say, they may be very closely linked with one another (as well as with Ag and L), each with an allelomorph producing an antigen not yet detected by an antibody. In this event, they are to be represented as Hu^A and Hu, He^A and He. Alternatively, they may be at the same locus, with a third allelomorph whose antigen production has not yet been detected: such genes would constitute the series Hu^{Hu} Hu^{He}, Hu (here Hu rather than He is arbitrarily chosen as the locus symbol, since Hunter was the earlier of the two to be discovered). The first of these alternatives, assuming separate loci for the control of these groups, is the more probable.

Both antibodies were produced by immunizing rabbits with Negro blood. The Hu(A+) condition has been found in 7 per cent of American Negroes, and in 22 per cent of West African natives: He(A+) is widespread in Africa, but with a lower frequency (2·7 per cent of West Africans). There is evidence from their association with the other MNL groups that crossing-over between Hu and He, considered together, and the Ag and L loci takes place more often than between the two latter.

(ix) THE SEX-LINKED SYSTEM

GENES: Xg^A, Xg.
ANTIGENS: Xg(A+), Xg(A−).
ANTIBODY: anti-Xg(A).

One human blood group, which was discovered by Mann, et al. (1962), is sex-linked. The presence of the antigen Xg(A) is dominant to its absence and due to a gene Xg^A.[1] The action of its allele Xg has not yet been recognized by the discovery of an antibody. Race (1965) gives the frequency of those possessing the antigen as 64·4 per cent of men and 89·2 per cent of women in a sample of 3,000 northern Europeans. The corresponding gene-frequencies are $Xg^A = 0.657$ and $Xg = 0.343$. The antibody anti-Xg(A) is exceedingly rare and when found seems to have arisen as an immune reaction.

The existence of a sex-linked blood group polymorphism, in which even the rarer phase is at a considerable frequency, will prove extremely valuable for mapping the X-chromosome, although the extreme rarity of the antibody is at present a handicap. The order and map-distance of the genes for this and three other sex-linked conditions have already been established with fair probability (Race, 1965), as follows: Sex-linked blood group, *23*, red colour-blindness, *4*, G6PD control, *5*, green colour-blindness, *12*, haemophilia. The figures in italics of course indicate cross-over values.

Gavin et al. (1964) have tested a number of mammals for the presence of Xg(A), including several Catarhine apes of the family Simiidae. They found it in 7 out of 13 gibbons, *Hylobates* (3 males and 4 females) but not among 67 chimpanzees, *Pan*, or 20 orangutans, *Pongo*, though these numbers do not exclude it as a polymorphism at a low frequency. In the gibbons, which are regarded as the most primitive of these three genera, it is indistinguishable, serologically at least, from the antigen occuring in man. It seems more likely that this blood group became established as a poly-

---

[1] Serologists have given this gene the notation Xg^a, which is standardized throughout genetics as that to be used for recessives. This is corrected here.

morphism in the common ancestor of man and of the gibbon, being lost, or reduced in frequency, in some other Simiidae, than that it has evolved independently in *Hylobates* and *Homo*.

(x) THE LEWIS SYSTEM

GENES: *Le*^{A}, *Le*.
ANTIGENS: Le(A+B—), Le(A—B+), Le(A—B—).
ANTIBODIES: anti-Le(A), anti-Le(B).

This system has been completely recast. Formerly it had given the impression of a blood group series controlled by three genes: one dominant in effect, one unrecognized by an antibody, and one recessive: *Le*^{B}, *Le*, *Le*^{a}. But it has now been discovered that it is not one primarily controlling the *blood* groups at all. The gene *Le*^{A}, dominant in effect as its notation indicates, is responsible for producing the Le(A) antigen in the saliva, and this its allelomorph *Le* prevents. The latter condition occurs in 9·8 per cent of Europeans and is recessive.

The Lewis and the secretor genes (pp. 152–3) of the O, A, B substances are unlinked but interact in their effects. Race and Sanger (1968) have, partly as an aid to memory, devised an explanation of this interaction. They suggest that there is a limited amount of water-soluble O, A, B antigens which can be made into A, B or Le(A) substance, the demands of the G^A and G^B genes (those controlling the A and B substances, pp. 147–8) being satisfied first. Thus in A, B secretors, some Le(A) antigen appears in the saliva but a negligible quantity reaches the plasma. In A, B non-secretors a large quantity of the Le(A) antigen can be formed. It then appears in bulk in the saliva and there is plenty to accumulate in the plasma also, from which it can be picked up by the red cells. Thus *serologically* only these are scored as Le(A): for it is now known that the Lewis antigen of the red cells is determined genetically, but that all the red cells can acquire this substance if it be available and, indeed, the superficial nature of its attachment is indicated by the fact that it can be washed off them.

Consequently individuals with Le(A) on their red cells are all non-secretors of the O, A, B substances. They amount to about

22 per cent of the population in England and simulate recessive inheritance since non-secretion is recessive (p. 152). The relevant antibody, anti-Le(A), is of course restricted to those who are genetically *LeLe*. In these it is uncommon, though it has now been found on many occasions.

A second Lewis antigen, Lewis-B, whose antibody has now been detected quite frequently, can also exist in the saliva, as well as in the plasma from which it also is acquired by the red cells; in fact it occurs on them in about 72 per cent of the English population. This situation is not yet fully understood. There does not appear to be a gene at the *Le* locus controlling the occurrence of the Lewis-B substance. Rather, this seems to arise as some interaction between the genes for the O, A, B groups, the secretor and the Le(A) genes. For all those who carry the Lewis-B antigen on their red cells are secretors of the O, A, B substances. That is to say, the interaction referred to determines whether Le(A+B−) or Le(A−B+) is the antigen present. To make the situation clear, it can be summarized by saying that the *LeLe* homozygotes account for all those who lack both of the Lewis antigens on their red cells: that is, those who are Le(A−B−). These can be either A, B, O secretors or non-secretors but they cannot secrete the Lewis antigen though all other types do so. In addition, two further features deserving notice are probably aspects of this fact. First, Le(A+B+) cells have never been found in any adult, though they have been detected in children of 18 months or less. Secondly, the behaviour of the Le(B) substance can be analysed without difficulty in individuals belonging to groups O and A2 of the A, B, O series (pp. 147–50); but some, though not all, anti-Le(B) sera agglutinate cells of the Le(B) type only if they belong to groups O, A2 or B, and not to A1, of the O, A, B series. There appears to be no good evidence that either of the Lewis antibodies has been responsible for haemolitic disease.

Several other rare, or so far little evaluated, blood groups or blood group systems have been reported.

CHAPTER 7

The Blood Group Polymorphisms:
Clinically More Important Types

7.1 THE ABO SYSTEM

GENES: G^A, G^B, G.
ANTIGENS: G(A—B—), G(A+B—), G(A—B+), G(A+B+).
ANTIBODIES: anti-G(A), anti-G(B).
GROUPS: O, A, B, AB.
(See also pp. 149–52.)

Cases frequently arise in which it is necessary to transfuse blood from one individual to another. It had long been realized that though such a procedure is sometimes highly successful it can frequently be attended with serious and often fatal results. The technique was, therefore, useless until these failures were explained by the discovery of blood groups in 1900. In that year, Landsteiner analysed the O, A and B types which, owing to the invariable presence of the naturally occurring antibodies whenever compatible with the antigens, are at once more easily detected and clinically more important than any of the other systems which have been found subsequently, especially those discussed in the previous chapter.

Landsteiner described the G(A) and G(B) antigens (though, of course, this notation was not then used), and found that the corpuscles may carry both, one or the other alone, or neither of them. Individuals, therefore, may be placed in any one of four groups, known as AB, A, B and O, depending on their content of these substances. In group AB, both antigens are present, and in O, both are absent.

Since anyone may give blood who does not carry the antigens corresponding to the antibodies in the plasma of the recipient, the possibilities of transfusion can easily be determined. They are represented in Table 2.

147

Group O forms the class known as 'universal donors', since their blood is compatible with all others of the ABO system (Table 2, first horizontal line within the table). However, the blood of all groups but their own is incompatible and fatal to them when they are recipients (ibid., first vertical line). On the other hand, AB can supply blood for no group but its own (ibid., lowest horizontal line), but can receive blood from any group (ibid., last vertical line).

TABLE 2. The possibilities of blood transfusion (T = transfusion permissible)

		Recipient			
		O	A	B	AB
	O	T	T	T	T
	A	—	T	—	T
Donor	B	—	—	T	T
	AB	—	—	—	T

The ABO group of any individual can be determined by testing the reaction of his erythrocytes with anti-G(A) and anti-G(B) sera. Thus transfusions can now be performed with safety. It is important not to use blood whose corpuscles will be agglutinated by the plasma of the recipient. The reverse procedure, that of injecting blood whose plasma can agglutinate the recipient's corpuscles, is generally harmless; for in these circumstances, the donor's plasma is too much diluted to produce detectable effects. It is for this reason that universal donors can give blood to groups A, B and AB as well as to their own group, O (Table 2).

The percentage-frequencies of the four ABO blood groups in southern England are: O, 43·5; A, 44·7; B, 8·6; AB, 3·2 (data from Ikin et al., 1939, based upon a total of 3,449 individuals).

These blood groups are controlled genetically by a system of multiple allelomorphs which is linked with the nail-patella syndrome, a heterozygous condition characterized by abnormal finger-nails and absence of the patella. The cross-over value between these two loci is 12 per cent. Group O is due to the operation of a pair of allelomorphic genes GG, recessive in effect. The groups A and B are each produced by a different gene substitution (G^A or G^B) at the locus of G. Both the A and B groups are dominant to group O. That is to say, the genotypes G^AG^A, and G^AG both give rise to group A and are not distinguishable in their effect; nor are G^BG^B and G^BG, producing group B. However, the genes G^A and G^B interact, resulting in the AB group which, genetically, is G^AG^B (as indicated by the notation).

The percentage of loci occupied by these three genes in southern England, resulting in the frequencies of the four ABO blood groups given above is: G, 66; G^A, 27·9; G^B, 6·1. Their distribution in other populations and races is discussed in the section on Anthropology.

It will be evident that, from the point of view of blood grouping, the results of any marriage can be predicted within certain limits. These can readily be deduced from the information just provided, and they are listed in Table 3.

The reaction of the antigen G(A) to its antibody may be either strong (group A1) or weak (group A2). The difference is not believed to effect the safety of blood transfusions, so that it is

TABLE 3. Ten types of marriage are possible with respect to the four blood groups O, A, B and AB. These are listed in the first column. In the second column will be found the blood groups which cannot appear among their children

Type of marriage	Blood groups absent among children
O × O	A, B, AB
O × A	B, AB
O × B	A, AB
A × A	B, AB
B × B	A, AB
A × B	—
O × AB	O, AB
A × AB	O
B × AB	O
AB × AB	O

not usual to discriminate between these types in ordinary blood tests; consequently Table 2 has not been subdivided in respect of them.

Genetically, groups A1 and A2 are controlled by two genes, G^{A1} and G^{A2}, allelomorphic to one another and to G^B and G. They form a multiple allelomorphic series in the order G^{A1}, G^{A2}, G^B, G, each of which is dominant in effect to those below it; except that G^B, though dominant to G, interacts with G^{A1} and G^{A2} to produce groups A1B and A2B respectively. It will be apparent, therefore, that group A1 may carry genes G^{A2} or G, but not G^B in single dose; but that groups A2 and B can carry only G in addition to the apparent gene.

The G^{A2} gene is almost entirely European and African (pp. 170, 176). In southern England it occupies approximately 7 per cent of available loci, out of a total of 27·9 per cent of G^A genes, so that 9·9 per cent of the population belong to group A2.

Remembering that not more than two of the genes of the ABO series can coexist in the same individual (if, indeed, they be multiple allelomorphs), the genetics of these blood groups, taking into account the subdivision of A in groups A1 and A2, can easily be deduced. This refinement naturally limits still further the possible results to be obtained from marriages of the various types. Such information, additional to that provided in Table 3, is given in Table 4.

Anti-G(A) is, in reality, a mixture of two antibodies: anti-G(A2) which reacts both with cells carrying the antigens G(A1) and G(A2), and anti-G(A1), which reacts only with those carrying G(A1). This is a fact which has been demonstrated serologically. It is most easily interpreted on the assumption that the G(A) antigen possessed by those of group A1 is itself a mixture of two antigens, as indicated in Table 5, in which the subdivision of group A is shown.

Several subdivisions of A, groups A3, A4 and A5 having still weaker reactions with anti-G(A), are now known. Presumably they are inherited as further allelomorphs at the G locus (or as other closely linked genes). These are all very rare and they do not seem to constitute a polymorphism at all; they are probably eliminated by selection and merely maintained in the population by mutation (p. 110). Consequently, they are of negligible importance in medical genetics and, though they may be of some interest in regard to the genetical theory of the blood groups

TABLE 4. Marriages involving group A, when this is subdivided into A1 and A2, and the blood groups which cannot appear among their children

Marriage	Blood groups absent among children	Marriage	Blood groups absent among children
A1 × O	B, A1B, A2B	A1B × A1	O, A2
A2 × O	A1, B, A1B, A2B	A1B × A2	O, A2, A1B
A1 × A1	B, A1B, A2B	A2B × A1	O
A2 × A2	A1, B, A1B, A2B	A2B × A2	O, A1, A1B
A1 × A2	B, A1B, A2B	A1B × B	O, A2, A2B
A1 × B	—	A2B × B	O, A1, A1B
A2 × B	A1, A1B	A1B × A1B	O, A2, A2B
A1B × O	O, A2, A1B, A2B	A2B × A2B	O, A1, A1B
A2B × O	O, A1, A1B, A2B	A1B × A2B	O, A2

(see Race and Sanger, 1968), they cannot be discussed further in the limited space here available.

Antibodies, known as anti-O and anti-H, react with an antigen manufactured by the gene *G*. This, however, does not fully explain that situation which, in fact, has not yet been entirely elucidated.

TABLE 5

Group	Antigens	Compatible antibodies	Incompatible antibodies
A1	G(A1+A2+B−)	anti-G(B)	anti-G(A1) anti-G(A2)
A2	G(A1−A2+B−)	anti-G(A1) anti-G(B)	anti-G(A2)
A1B	G(A1+A2+B+)	—	anti-G(A1) anti-G(A2) anti-G(B)
A2B	G(A1−A2+B+)	anti-G(A1)	anti-G(A2) anti-G(B)

The antibodies in question are found in the sera of some humans and many other mammals, chicken, eels and in certain seeds. All that can be said here is that they react most strongly with cells of groups O and A2, and that anti-H is inhibited by the addition of secretor saliva of any group (see the next section) while the anti-O type is not. Donors of anti-H are of the Le(A+) type while those of anti-O are normally distributed in relation to the Lewis groups.

It has already been pointed out that anti-Ag(N) is present in the seeds of *Vicia graminea*. Antibodies of the O, A, B series have also been found in certain seeds (Mäkelä, 1957). For instance, an extract of *Vicia cracca* agglutinates cells of group A (A1 more strongly than A2) but of no other groups, while an extract of the seeds of *Cytisus sessilifolius* agglutinates only cells of groups O and A2. The lima bean, *Phaseolus limensis*, and the related *P. lunatus*, are also powerful sources of anti-G(A). Moreover, the seeds of *Ulex europaeus* contain a large amount of anti-G(H), which is of use in distinguishing secretors and non-secretors of the O, A, B antigens (see next section). *Dolicos biflorus* agglutinates A1 cells and also those carrying Sd(A). The only non-Papilionaceous plant in which the A antibody (A1) has been found is *Hyptis suaveoleus*, Labiatiae. The significance of the presence of these antibodies in plants is still in doubt.

7.2 THE ANTIGEN SECRETION OF THE ABO SYSTEM

The antigens G(A) and G(B) are often found in various gland cells and in the secretions which these produce. Thus they may be present in small quantities in tears, urine and certain other body fluids (though not, apparently, in cerebrospinal fluid); while in saliva and semen they can occur in high concentrations. This is due to the action of a gene *S*, dominant in effect, which allows their conversion to a water-soluble form. These antigens are always alcohol-soluble on the erythrocytes, and they are confined to that state by the genotype *ss*, which is recessive in effect. Therefore they cannot be secreted by such individuals.

The classification of secretors and non-secretors is not always clear-cut, and considerable technical difficulties are involved in it. However, Race and Sanger (1968) consider that just over 22 per cent of Europeans are non-secretors. The frequency varies in different races, being lowest in North American Indians. It will be realized, therefore, that these 'secretor' genes are maintained as a polymorphism.

There is linkage between S and Lu, the locus of the Lutheran genes, C.O.V. = 9 (Mohr, 1951). It was until recently supposed that secretor and Lewis were also linked. This is not so, the association between them being due to their interaction described on pp. 145–6. In addition to the A and B antigens, Le(A) and Le(B), as well as Sd, but no other antigens, are present in the saliva.

7.3 THE RHESUS SYSTEM

GENES: C^A, C^B; D^A, D; E^A, E^B (and see p. 157).
ANTIGENS: C(A+B+), D(A+), E(A−B+) could represent
(examples   the antigens of one donor.
only)   Individual antigens could be stated as E(A+), or
              (A−).
ANTIBODIES: anti-C(A), anti-C(B), anti-D(A), anti-E(A), anti-
              E(B).

Only in the ABO series and anti-Tj(A) are the naturally occurring antibodies invariably present in compatible blood. However, in some of the Rhesus groups the immune antibodies are produced so easily that they are not infrequently responsible for serious haemolytic disturbances.

The human race is usually classified as Rhesus-positive or Rhesus-negative, depending upon the antigens belonging to this system which are present upon the erythrocytes (pp. 158–60). These are inherited as nearly complete dominants. Their antibodies are rare, though not unknown, when spontaneous. However, they were originally obtained by injecting rabbits and guinea-pigs with the blood of Rhesus monkeys, whence the name, and it was found that the sera so produced could be used for detecting corresponding antigens in human blood. About 84 to 86 per cent of the population are of the type known as Rhesus-positive in most European countries; that is to say, they carry one or more of the

Rhesus antigens which frequently, or not very infrequently, stimulate immune antibodies. Racial differences, however, exist (pp. 166–75); for instance, the Rhesus-negative condition is very uncommon among the Chinese, apparently less than one per cent.

Serious reactions are sometimes obtained when a person of either sex, who has been given a successful injection of group O blood (the universal donor of the ABO system), is transfused a second time. A similar misfortune may occur when women receive their first transfusion (with group O) if this follows the birth of a child, or a miscarriage. Incompatibility of the Rhesus groups is now known to be responsible for the majority of these calamities which, until the discovery of Rhesus in 1940, were a considerable menace to blood transfusion.

As already explained, Rhesus antibodies are not normally found in human blood, but it is now clear that they may be formed if a group O injection containing a Rhesus antigen, particularly D(A), be given to a Rhesus-negative person. These antibodies may then persist almost indefinitely in the plasma. When, therefore, such a Rhesus-negative person receives a second injection of Rhesus-positive blood, interaction between antigen and antibody can produce grave results.[1]

It is possible for a Rhesus negative woman to give birth to a Rhesus positive child (or miscarriage), whose ability to form the antigen has been inherited from its father. If there is a foeto-maternal haemorrhage at or near delivery, red blood cells escape into the mother's circulation, and after some months she may form Rh antibodies. These may cause haemolytic disease of the newborn in a subsequent Rh positive pregnancy, since the antibody is of the 7S type which can cross the placenta. The accumulated pigment from the destroyed erythrocytes leads to jaundice and sometimes brain damage (due to kernicterus).

A naturally occurring protective mechanism against Rh immunization is ABO incompatibility between mother and foetus (e.g.

[1] The following four paragraphs have most kindly been drafted for the present edition of this book by Professor C. A. Clarke, F.R.S., who is himself responsible for devising the method of preventing Rh immunization of a mother which he here describes (see p. xvi).

mother O, foetus A), for here the Rh positive foetal red cells are destroyed by the mother's naturally occurring anti-A or anti-B. This is only one of several protective mechanisms, the other most important one being that about 30 per cent of people are incapable of making Rhesus antibodies, no matter how big the stimulus. The net result is that the frequency of Rh haemolytic disease of the new-born (only 1 in 150 or 200 infants) is much less than would be expected from the mating types (in most European countries the father is Rhesus positive and the mother Rhesus negative in about 1 marriage in 8). Rh haemolytic disease is also responsible for some of the stillbirths which take place after the 28th week, but seldom for those occurring earlier.

Where the husband is homozygous, as three out of seven Rhesus positive men will be, the chance of bearing a healthy baby once one suffering from haemolytic symptoms has appeared is remote. Exchange or intrauterine transfusion, however, gives good results, though there are dangers – associated with prematurity – since the baby is often induced before term in these cases.

These methods of treating the established disease are now much less used because of a very effective method of *prevention* of Rh immunization of the mother. This is by giving the mother an injection of 100 μg of 7S anti-Rh gammaglobulin within 72 hours of delivery. This coats any Rh positive antigen sites on the foetal red cells and renders them non-antigenic. It is usually considered unnecessary to give the injection where the baby is incompatible with the mother on the ABO system. The standard dose is effective for 98 per cent of cases but where the foeto-maternal haemorrhage has been very large more must be given. The same is true when a mismatched transfusion has been administered, though this is now a very rare event since all Rh negative individuals are routinely given Rh negative blood. Evidence from all parts of the world where Rh haemolytic disease is a problem shows that the degree of protection is very high, the failure rate being between 1 and 2 per cent. This has to be compared with a 17 per cent immunization rate if nothing is done.

The Rhesus series then is a composite one, of which the anti-D(A), anti-C(A) and anti-E(A) antibodies were the first to be

155

obtained. They, of course, recognize their corresponding antigens, D(A), C(A) and E(A) which, singly or in combination, produce the Rhesus-positive condition (see pp. 158–9). These antigens are each unifactorily controlled by the genes D^A, C^A and E^A, which are nearly dominant in effect. Such groups of genes having extremely similar effects could arise in a number of different ways. They might be intracistronic, or they might be major genes lying very close to one another: in that event, they could have been pulled out of the same cistron, and indeed Parsons (1958) has been able to increase cross-over values experimentally. Alternatively, they might be super-genes, brought together from different parts of the same chromosome or from different chromosomes because of their similarity of action and the need to combine them into a group. In Rhesus, there seems evidence that the C and E loci are intracistronic. For anti-C(B) and anti-E(B) react with red cells doubly heterozygous for $C^A C^B$, $E^A E^B$ when in the *cis* but not in the *trans* relationship. Whether or not D is included in the cistron with them seems yet uncertain, but at any rate it does not lie between C and E.

Dr R. R. Race informs me that what appeared to be another antigen F(A), determined by a gene F^A, is really an interaction between the genes C^B and E^B.

These three most important Rhesus antigens are of unequal frequencies, though all are quite common; the percentages of available loci occupied by the genes controlling them being D^A, 59; C^A, 42; E^A, 15·5.

Antibodies, anti-C(B) and anti-E(B), recognizing the C(B+) and E(B+) conditions have now been discovered. These antigens are produced by the allelomorphs of C^A and E^A; that is to say, by the genes C^B and E^B. They are dominants but with a slight dosage effect, so that the agglutination of the homozygotes is a little stronger than that of the heterozygotes. An antibody detecting the antigen situation D(B+) has been reported, but its existence does not seem fully substantiated and it is best to await confirmation of it. Consequently, the allelomorph of the gene D^A must, at present, be written D. When its corresponding antibody (and, therefore, its antigen) is obtained, as presumably it will be, this gene will become D^B if dominant in effect, as is to be expected, but D^b if recessive.

Other Rhesus genes have been discovered. These are rare, and they appear to be additional allelomorphs at the C, D and E loci. Of these, C^W, producing the C(W) antigen detected by its antibody anti-C(W), is best known. It is approximately dominant and, since it occupies as much as 1·3 per cent of the C loci, it also must be a polymorphism. Further Rhesus genes, D^{U1}, C^U, C^X, E^U, E^W and others have been detected.

D^{U1} has been studied in some detail. It occupies a rather small proportion of D loci, not yet precisely ascertained. It can be recognized in the homozygous and heterozygous conditions, $D^{U1} D^{U1}$ and $D^{U1} D$, but not in the $D^{U1} D^{A1}$ genotype, as indicated by the introduction of numbers into the notation: for a large figure in the suffix represents a gene dominant in effect to one with a small figure, so that D^{A1} is dominant to D^{U1}, and both are dominant to D (and it has been explained, pp. 132–3, that a capital in the suffix indicates dominance over no letter in the suffix). We have here the same notation as that employed for the G^{A1}, G^{A2}, G genes of the ABO series. These numbers should, therefore, be added when D^U is being discussed, otherwise the number can be omitted from D^A.

Two distinct 'shorthand' notations have been developed in research laboratories to indicate the different combinations of genes at the D, C and E loci (those at the supposed F locus were not included). These are for specialized use within the field of blood grouping only. They should not be used except by those familiar with the genetic and serological situations as defined by the full notation, for they are mere symbols chosen to represent a selection of the Rhesus genotypes and phenotypes. However, they save much space when extensive lists are compiled, such as those constructed for anthropological work. The two shorthand systems are sufficiently alike to be confusing, and that used in England (see Race and Sanger, 1968) should always be adopted, for it has needed no revision. The alternative convention, due to Wiener (1949) is to be avoided, for it has been confused by repeated alterations in respect of the phenotypes.

The English system is shown in Table 6, which also gives the frequencies of the different arrangements in Britain. It is based, with kind permission, upon a table provided by Race, Mourant, Lawler and Sanger (1948). However, the modern gene-notation is substituted for that then in use, and the frequencies of the various chromosome types are expressed in percentages, instead of in fractions of unity: See also Race and Sanger (1968).

TABLE 6. The frequencies of the Rhesus chromosome-types in England. (Based upon Race, Mourant, Lawler and Sanger, 1948.)

Genes	Abbreviated notation	Frequency per chromosome %
$D^A C^A E^B$	R_1	40·76
$D\ C^B E^B$	r	38·86
$D^A C^B E^A$	R_2	14·11
$D^A C^B E^B$	R_0	2·57
$D^A C^W E^B$	R_1^W	1·29
$D\ C^B E^A$	R^{11}	1·19
$D\ C^A E^B$	R^1	0·98
$D^A C^A E^A$	R^2	0·24
$D\ C^W E^B$	R^{1W}	
$D\ C^A E^A$	R_Y	Very rare
$D^A C^W E^A$	R_Z^W	
$D\ C^W E^A$	R_Y^W	

It will be noticed that these frequencies represent the occurrence of single chromosomes, and from them the individual genotypes can be calculated. Thus the Rhesus-negative condition $(D\ C^B E^B)/(D\ C^B E^B)$ occurs in 38·86 per cent of individuals, which equals 15·1 per cent of the population.

Having regard to the genetical and serological conditions of Rhesus as so far explained, the concept of the extremely important 'Rhesus-positive' and 'Rhesus-negative' states will appear confusing. It is, in fact, fundamental for practical serology but, theoretically, illogical. All the allelomorphs at the Rhesus loci are known to produce their appropriate antigens (except for D; and, as already indicated, we may feel certain that the one due to this gene has not yet been, but will be, detected rather than that it is non-existent). Thus, in one sense, all the Rhesus allelomorphs are capable of producing haemolytic reactions and, therefore, of acting as Rhesus-positives. The Rhesus-negative concept depends upon the fact that many of them very seldom do so because their antibodies, whether immune or spontaneous, are formed only with great rarity. Consequently, they are no more, but perhaps no less, a menace to blood transfusion, and a source of haemolytic disease in the new-born, than are the blood groups discussed in Chapter 6.

In fact, only D^A, in either genotype, or homozygous C^A or E^A, sufficiently often stimulate immune antibody formation to be classed unequivocally as Rhesus-positive. Heterozygous C^A or E^A individuals, that is to say, those of the constitution $(D\ C^A E^B)/$ $(D\ C^B E^B)$ and $(D\ C^B E^A)/(D\ C^B E^B)$, are usually treated as Rhesus-positive when donors, but Rhesus-negative when recipients. Moreover, anti-D(A), arising in those of the $D\ D$ genotype, is so much the most frequent of the antibodies that it is responsible for about 95 per cent of all instances in which they occur. It might almost be said, therefore, that effective Rhesus-positives are limited to the D(A+) condition. In Britain about 15 per cent of the population are of the genetic constitution $(D\ C^B E^B)/(D\ C^B E^B)$ and these are the definitive Rhesus-negatives (p. 165). With them other types can evidently be included, for example $(D\ C^W E^B)$, when homozygous or heterozygous for C^W, but these are so rare that they do not materially influence the frequencies.

Since in England the husband is homozygous Rhesus-positive and the wife Rhesus-negative in about one marriage in nineteen, while less than one newborn child in 150 suffers from haemolytic jaundice, it is clear that by no means all those infants who are at risk do in fact develop the disease (p. 154). Furthermore, Clarke (1963) found great variability in the immune response to the D-antigen by volunteers, both male and female, when injected with Rhesus-positive blood. Professor C. A. Clarke has been so good as to inform me that as many as 30 per cent of people make no Rh antibodies no matter how much or how often they are given Rh+ blood. He remarks that it seems likely that they fail to respond to other red cell antigens as well. Evidently some protective mechanism is at work here, and Clarke (1971b) makes the stimulating suggestion that ability to form the antibody may be partly inherited; if so, haemolytic disease should run in families. As he points out, this could be tested by studying the Rhesus-negative sisters of immunized women and comparing the rate of their immunization with that in sisters of non-immunized controls.

The Rhesus loci are linked with the gene, dominant in effect, which produces elliptocytosis; that is to say, oval red corpuscles, a condition normal in the Camelidae. It is nearly symptomless, but may produce anaemia when homozygous. The C.O.V. has been estimated as 10–15 per cent (p. 182).

Some information on the occurrence of the Rhesus genes and super-genes in different races is given in the section on Anthropology.

The various polymorphisms described in this and the previous chapter are of course under independent control but we should expect to find inter-relations between them, and at two levels. First, that of the physical basis of heredity; for we might anticipate that some of the major genes involved might be linked, as in fact those for the secretor condition and the Lutheran groups have proved to be. Secondly, it is reasonable to expect physiological interaction between polymorphisms affecting serological characters. Of this the inter-relation between Lewis, A, B, O and secretor (p. 145) provides an example. A further instance of the kind remains to be mentioned. That is to say, the incompatible children of individuals incompatibly mated on the O, A, B system are better protected than those of compatibles against haemolitic disease of the newborn when due at least to the Rhesus and the Kell groups, and probably to other groups as well. It will be apparent that we have here a situation which favours diversity and tends to maintain the polymorphisms involved (Clarke, 1971b).

It is to be noticed that the ABO blood groups can themselves lead to haemolytic jaundice in the newborn (Clarke, 1971b, pp. 324–6). It has been pointed out (p. 155) that not all cases at risk develop the form of this disease based upon the Rhesus groups. So it is with that due to ABO, which in England is present only in about 3 per cent of infants, where about 20 per cent of births involve incompatibility between mother and child in respect of this blood group series. Therefore protective mechanisms must be operating here also.

Group O mothers are almost the only ones concerned because they, more readily than groups A or B, form a type of anti-A and anti-B with a small enough molecule to cross the placenta. This could then affect the foetus when it belongs to group A1 (rarely A2) or B. But even when these antibodies do cross the placenta, they are generally rendered innocuous by the blood group antigens in the plasma, perhaps particularly in Secretors, so that the red cells are not damaged. This, Professor C. A. Clarke points out, is the principal natural protective mechanism. In contrast with the Rh type, it is generally the first child, incompatible relative to ABO, which shows the disease; certainly in some instances because the mother has been immunized by some non-human antigen (carried in horse-serum, for instance). Unlike Rhesus haemolytic jaundice, that due to ABO is generally mild and disappears spontaneously without doing permanent damage.

CHAPTER 8

Further Aspects of the
Blood Groups

8.1 CLINICAL EFFECTS AND POLYMORPHISM

The blood groups were for the first time treated as examples of balanced polymorphism in the original edition of this book (1942), when it was therefore pointed out that they are not of neutral survival-value: a fact already deduced by Ford (1940*a*). At that time, this approach to the subject was by no means universally approved. In the interval, however, it has been completely justified on the basis of direct evidence, quite apart from the considerations outlined in Chapter 5, which established it on theoretical grounds.

Such direct evidence is chiefly available for ABO and Rhesus now, though it will, doubtless, be extended to the other series in the future. The polymorphism of the blood groups may be maintained either by other effects of the genes concerned, in addition to their control of antigen and antibody formation, or as a result of a serological interaction and corresponding adjustment between mother and foetus.

Considering that the major genes so constantly have multiple effects, and in view of the theoretical necessity that those of the blood groups are maintained in the population by balanced selective advantages, it had long seemed probable that some serological types are more prone to develop certain diseases than are others. It was, indeed, disappointing that for many years no attempts were made to compare the blood groups of, for instance, patients suffering from tuberculosis, or cancer at particular sites, with a random sample of the population to which they belong. This work has now begun, and is already giving results of importance.

Aird, Bentall and Frazer Roberts (1953), have shown that a significantly high proportion of those suffering from cancer of the stomach belong to group A of the ABO series. Moreover, Clarke *et al.* (1955) do not find an excess above expectation of A secretors, compared with A non-secretors, among them. Consequently, the carcinogenic stimulus is more likely to be due to some other effect of the gene G^A, than to direct irritation of the stomach wall by large quantities of the G(A) antigen swallowed with the saliva.

Clarke *et al.* (*l.c.*) show that duodenal ulcer is commoner by about 30 per cent in individuals of group O than in those of the other OAB blood groups and commoner still, by about 40 per cent, in those who fail to secrete the G(A) and G(B) antigens in their saliva. Gastric ulcers are also much more frequent in group O individuals than in others, while pernicious anaemia is found disproportionately often in group A. Pike and Dickens (1954) have demonstrated that the O blood group is a predisposing influence in the production of toxemia of pregnancy, while the other ABO blood groups are at an advantage in this respect (53·6 per cent of the patients belong to group O, compared with 45·5 per cent of the controls). There seems also to be good evidence for a deficit of blood group O among those with ischaemic heart disease (Allan and Dawson, 1968). Infections were included in the association between the human blood groups and disease when this was originally suggested, and it now appears that those who respond to type-A haemolytic streptococcal infection by developing rheumatic fever are significantly deficient in blood group O and in secretor (Clarke, 1964, p. 181).

Moreover, it must have been a matter of much importance in the evolution of the human races that individuals belonging to blood groups A and AB are much more liable to develop smallpox, and to do so in a more severe form, than those belonging to groups O and B. This was denied on experimental grounds by Harris *et al.* (1963) and others when first reported, but the association has now been fully established by Vogel and Chakravartti (1966) working in Western Bengal and Bihar, and confirmed by Bernhard (1966) both in India and Pakistan. In the 'natural situation' the effect seems to be limited to severe epidemics in unvaccinated populations.

Perhaps it is for this reason that it has taken so long to detect and confirm it, although when the necessary connection between blood groups and disease was originally pointed out in 1945, infections were specifically included. That prediction was reached by applying the concepts of ecological genetics to medicine (Ford, 1971). Extending that point of view, it may be said that the balanced gene-complex against which the blood groups, and indeed all polymorphisms, of any population must be adjusted, may influence the relationship between them and other conditions. Thus although the association between the OAB blood groups and smallpox that has now been detected in Indian peoples is a strong one, the correlation may hold to a different degree in other parts of the world.

A number of further conditions have, with considerable probability, been associated with certain serological types. Work on these is proceeding. There are always to be found those who wish to minimize the importance of selection. It chanced that when the relation between the blood groups and disease was originally established after its prediction, the pathologies involved were those associated with the alimentary canal. Accordingly it was suggested that these were the only types to be correlated with serology. That view could not be maintained now, nor indeed was any such restriction likely on theoretical grounds; while, indeed, it had already become untenable in view of the association between pernicious anaemia and blood group A.

More general physiological effects of the blood group genes upon the population as a whole have now been detected, in addition to such specific correlations with particular diseases so far mentioned (Allan, 1954). Thus there is clear evidence of selective elimination operating against OAB incompatible foetuses: for instance, a group A foetus in a group O mother (Matsunaga, 1962). About 10 per cent of such zygotes seem to be lost differentially in European and American families, and 3 to 21 per cent in the Japanese population; the variability in the latter being associated with environment, ranging from favourable, with good medical attention, to unfavourable in which the higher eliminations are recorded.

As we might expect, heterozygous advantage has evolved in

the blood group polymorphisms. This is particularly clear in respect of AB × AB matings, in which the genotype of all the offspring is certain, and Chung and Morton (1961) find that they give rise to a heavily significant excess of AB children; a situation which will doubtless become clear in groups A and B, as for Rhesus, also when the two genotypes of each can be distinguished. Similar heterozygous advantage has also been established in respect of M, N blood groups (Clarke, 1964, pp. 78–79).

It will be noticed that the selective effect of the association between blood groups and disease is likely to be extremely small. It is, moreover, non-existent in such a condition as cancer of the stomach which is almost confined to those above the age of reproduction, though this may well not have been so at an earlier stage in evolution. However, as with ability to taste phenyl-thio-urea (p. 124), such associations show that the genes controlling the blood groups have effects additional to their serological ones; and some of these may be of much importance to the individual, as in the elimination of incompatible foetuses and the existence of heterozygous advantage. Such conditions are in accord with the existence of the balanced selective forces which must be operating upon these, as upon other, polymorphisms.

It will have been noticed in the discussion of Rhesus in Chapter 7, that selection discriminates against the offspring of marriages between Rhesus-positive men and Rhesus-negative women. This union tends to produce babies suffering from haemolytic jaundice at birth, from which a considerable proportion (approximately 50 per cent) die if untreated. This must tend to reduce the frequency of the Rhesus heterozygotes, especially of the type $D^A D$, since this locus is the one most responsible for antibody production. The effect must be towards eliminating the rarer of the allelomorphs concerned, especially, therefore, D. We may feel confident that no such elimination is taking place, and that the disadvantage of D is somehow counter-balanced. It has been suggested by Glass (1950) that this is partly due to a tendency to compensate for these deaths by adding more children to the family, for not all affected babies die. Such an adjustment cannot, however, be effective in those societies in which family limitation is not practised, and it is probable, therefore, that (as may be anticipated, p. 165) the Rhesus-positive heterozygotes are at some physiological advantage in regard to general viability, fertility, or protection from disease, of the kind to which attention has just been drawn.

Turner (1969) has shown the double homozygotes[1] of the MNL groups to be at a marked selective disadvantage compared with single homozygotes. As he says, the most fit will probably be double or single heterozygotes, a fact already known for MN (p. 165).

In view of such findings as these (and see p. 162 *et seq.*), it is now impossible to maintain that the blood groups are of neutral survival value, as has been done by Wright (1940), and others. On the contrary, the selective importance of the blood groups is now being established for a variety of conditions.

Since genes have multiple effects, it is to be expected that those of the blood groups may control also a variety of minor features some of which could be detected by a close comparative study of races with distinct serological qualities. One instance of this kind has been recognized which is of great interest in regard to anthropology and the evolution of language. Darlington (1947) shows that there is in Europe a remarkable agreement between the people who use the ð and þ sounds and those possessing 64·5 per cent or more of the gene G of the ABO series, except for the Italian peninsula and the islands between it and Spain.

8.2 THE BLOOD GROUPS IN ANTHROPOLOGY

The distribution of the blood groups, poised as these are in equilibrium by contending selective advantages and disadvantages, provides one of the most important criteria of anthropology. A large body of relevant information has already accumulated for ABO, MN and Rhesus in certain parts of the world. Elsewhere the data are scanty, or wholly lacking (as, for instance, in north-eastern Siberia), while the other blood group series have, as yet, been less fully exploited from this point of view. However, from what has

[1] Ag^ML^A/Ag^ML^A, Ag^ML^B/Ag^ML^B, Ag^NL^A/Ag^NL^A, Ag^NL^B/Ag^NL^B.

already been said of them in Chapter 6, it will be clear that they offer great possibilities for such studies which, in some respects, have already been realized.

We are fortunate in possessing an accurate and extremely thorough survey of the distribution of the human blood groups by the chief authority on the subject, Mourant (1954). To that account, I am principally indebted for the information contained in this section.

Comparative studies of the blood groups have been carried out more extensively in north-western and central Europe than elsewhere. Here the frequency of the gene G^A nearly always lies between 25 and 30 per cent of available loci, so that approximately 44 to 50 per cent of the population belong to the A and AB groups combined. This is a high frequency compared with the rest of the world. It is, indeed, characteristic of the peoples who, since late pre-historic times, have occupied Europe. As already explained, G^{A2}, the gene frequency per cent of which is approximately 4 to 8 (or about 1 in 4 to 1 in 6 of all G^A genes), is almost entirely European and African. G^B is of low frequency (5 to 8 per cent of loci) in northern and central Europe. The Rhesus genes are here very uniformly distributed, and so are those of the MN group. Indeed, as Mourant points out, such uniformity is one of the more striking characteristics of this region.

These populations, however, include two exceptional races, which must be mentioned separately. The Lapps are, as might have been expected, to some extent Asiatic in their blood group distributions: in their very high frequency of the Fy^A allelomorphs of the Duffy series, in some Rhesus resemblances and, furthermore, in a high proportion of those who can taste phenyl-thio-urea. In other respects, however, Lapps do not at all conform with what is known of Asiatic peoples. Thus in them the $(Ag^N L^A)$ supergene reaches one of the highest frequencies in the world while, in their possession of the G^{A2} gene at 30 per cent of available loci, the Lapps are apparently most un-Asiatic. However, we have little information on the blood groups of Siberia, and comment on the Lapp situation should await such data.

The Basques are the other entirely exceptional race to which reference has been made. They probably represent a survival of the 'pre-European', that is to say, late Palaeolithic inhabitants of the Continent. This is indicated by their osteological features, their region of distribution, and the fact that they, alone among the peoples of western Europe, do not speak an Indo-European language.[1] This contention is certainly supported by their blood groups, so unlike those of the peoples by whom they are encircled. They have the lowest frequency of the G^B gene in Europe, varying from 0 to 3 per cent of loci in different surveys, while D of the Rhesus series is exceptionally common among them (55 per cent compared with 40 per cent elsewhere in the region). In general, the Basques, therefore, contain a high proportion of Rhesus-negatives.

The far north-western lands of Wales, northern England, Scotland, Ireland and Iceland, are characterized by a markedly higher frequency of the gene G of the ABO series (70 to 75 per cent or more) than that found in southern England or Continental Europe (70 per cent or less). The frequency of G^B is also very slightly, but consistently, high, while that of G^A is significantly low. The sharp demarcation of this district of high frequency of group O is in accord with the facts of anthropology and history: that it has consistently been a region relatively undisturbed by immigration from the East, and into which the inhabitants of less remote parts have been driven by such invasions.

Eastern Europe is characterized by a greater frequency of the G^B gene than is western, exceeding 10 per cent. Indeed, except that it includes the eastern half of Germany, the boundary of this area agrees well with the western limit of Slavonic languages. The frequency of G^B rises as we pass towards south-eastern Asia, where it reaches the highest known value (see Table 7). Mourant makes an important summary of the situation when he says: 'The most striking general feature of eastern Europe as a whole is the presence of a series of concentric tongues of high B and low O frequency, based on the northern end of the Caspian Sea and sweeping across south Russia and Poland into eastern Germany. Their form sug-

[1] It is not clear that the Lapps provide a genuine exception to this statement, since their language may be obtained secondarily from Finnish.

gests that they mark the track of extensive immigration into central Europe from the East.' Neither the MN nor Rhesus genes differ to any considerable extent from their occurrence in the western districts.

TABLE 7. Percentage frequencies of the blood groups of the O, A, B series in different races (arranged in increasing frequency for group B)[1]

Race	Total examined	Frequency (per cent) of blood groups			
		O	A	B	AB
Australian (aborig.)	603	54·3	40·9	3·8	1·0
Dutch	14,483	46·3	42·1	8·5	3·1
English (southern)	3,449	43·5	44·7	8·6	3·2
Dutch Jews	705	42·6	39·4	13·4	4·5
Russian Jews	1,475	36·6	41·7	15·5	6·1
Bushmen	336	83·0	—	17·0	—
Hungarians	1,041	29·9	45·2	17·0	7·9
Arabs	2,917	44·0	33·0	17·7	4·1
Japanese	24,672	31·1	36·7	22·7	9·5
Russians	57,122	32·9	35·6	23·2	8·1
Negroes (Congo)	500	45·6	22·2	24·2	8·0
Chinese (Canton)	500	45·5	22·6	25·0	6·1
Hungarian gypsies	925	28·5	26·6	35·3	9·6
Hindus	2,357	30·2	24·5	37·2	8·1

[1] The English frequencies are from Ikin et al. (1939). The Hungarian and Hindu frequencies are from Backhausz et al. (1950) Homo, 1: those of other races are selected from the compilation of Wiener (1939). The Australian aborigines listed here are probably not pure-bred.

Here again one race, entirely exceptional on ethnological grounds, requires mention: that is to say, the Finns. They are to be compared with the Basques and Hungarians in speaking a non-Indo-European language. Like the Scandinavians and the Lapps, they have a very high frequency of G^A. Yet they are sharply distinguished from the Scandinavians in their much greater frequency of G^B, and from the Lapps in the rarity of their G^{A2} genes.

The populations bordering the northern shores of the Mediterranean tend to have a slightly lower frequency of the $(D\ C^B E^B)$

super-gene (r) than those of central and western Europe. This is most marked in Sardinia, where it occupies only 22 per cent of loci, compared with 38·9 per cent in England. The deficiency is principally counter-balanced by a greater proportion of $(D^A C^A E^B)$; that is, R_1.

In this northern Mediterranean region, there is considerable variation in the distribution of the ABO groups, but the main distinction of western and eastern European types (with high G^A and high G^B respectively) is applicable to them also. The 10 per cent level of G^B includes the eastern shores of the Adriatic, but passes north of Greece. A high G^B frequency continues down the Levant, through Egypt and, in general, along the North African coast. In the latter region, however, the Berbers of the Atlas Range are most exceptional. They have a very low frequency of G^A, down to 6·7 in one tribe, and a rather low G^B (reaching as little as 6·7), a low Ag^M frequency (18 per cent), and a very high occurrence of D, actually 55 per cent. They appear to be a re-markable anthropological group, perhaps with Basque connections.

Certain outstanding features characterize the Negro blood groups, of which a few only can be mentioned here. They com-bine high frequencies both of G, as in western Europe, and of G^B, as in the Near East, while G^A is relatively rare (10 to 20 per cent), being no commoner than G^B. As already mentioned, these are the only people, beside Europeans, who carry the G^{A2} gene, save as a rarity, which they do to about an equal extent. In the MNL series, L^A is shared rather evenly between the Ag^M and Ag^N genotypes, instead of preferentially accompanying Ag^M, as in Europe. The frequency of P^A seems universally high among Negroes. However, their most marked serological quality is the great preponderance of the $(D^A C^B E^B)$ super-gene (R_O, in the shorthand nomenclature, p. 158) which occupies 60 per cent of available loci, compared with 2 per cent in Europeans, and less than 15 per cent elsewhere in the world.

The frequency of the Fy^A gene is low; 6 per cent in Bantu, compared with 41 per cent in Europeans. There is, of course, much variation among the different Negro races, but these features are maintained among them. They are also characterized by the possession of the Hunter and Henshaw groups, nearly absent elsewhere (but see below). Hu(A+) individuals are found in 22 per cent of West African natives; their greater rarity (7 per cent) in American Negroes may in part be due to crossings with Europeans. The He(A+) condition is much rarer, about 2·7 per cent.

Asia is the region of highest G^B frequency, often over 25 per cent of available loci, giving proportions of the combined B and AB groups ranging, in general, between 35 and 50 per cent of the population. It is a surprising fact that this is equally true of the Caucasoid and Mongoloid peoples, indicating that the situation is palaeogenic (p. 180). The $(D\ C^B E^B)$ super-gene (r) occupies about 20 per cent of loci in south-west Asia and India. Further east, however, the D allelomorph is almost completely replaced by D^A, so that this is an area of overwhelmingly Rhesus-positive populations. Fy^A of the Duffy series seems universally very common among the Asiatic races, and so, on the whole, is Ag^M, but more so among Caucasians than Mongolians.

In Indonesia, the ratio of G^B diminishes eastwards from Java and Borneo, where it retains the high Asiatic value. The MNL system also relates Java and Borneo to Asia, for here are to be found high frequencies of Ag^M, while eastwards the Ag^N allelomorph predominates, becoming commoner than anywhere else in the world: 70 per cent of loci and upwards, to 91 per cent in Papua. $(D^A C^A E^B)$, that is to say, R_1 is here the characteristic super-gene of the Rhesus series, while the D allelomorph continues almost absent, as in Asia. The Lu(A+) condition seems nearly unknown here. It is a fact of great interest that among a sample of thirty-one Sea Dyaks tested, two were positive for the almost purely African Henshaw group, though a few He(A+) individuals have also been found in Asia.

Traces of a dark curly-haired people allied to the African Negroes have been found in India and Indonesia. It is to them also that the Melanesians belong. It seems possible, therefore, that negroids may once have extended from Africa through south India and out to Fiji, and it is perhaps from them that the He^A genes of Asia and Indonesia are derived.

The Australian aborigines are the most distinct and primitive race in the world; culturally they live in the Stone Age. Their outstanding serological features are twofold. First, G^B appears to be wholly absent (or, rather, reduced to the mutation level), so that the population is not polymorphic at all for this gene. It is, however, truly polymorphic for G^A and G: for G^A ranges from a gene-frequency of about 10 to 40 per cent, as a north to south cline. Secondly, L^A of the MNL system is also absent, though it occupies 20 per cent of the loci in New Guinea. In their other serological attributes, the Australians resemble the peoples of eastern Indonesia; for example, no Lu(A+) individuals have been found, while the Ag^N gene is exceptionally common, and D approximately absent, among them.

The vast differences between the various aboriginal inhabitants of the New World are, of course, reflected in their serology. Even so, as Mourant (1954) points out, certain general characteristics are shared by all of them, in North and South America alike. The frequency of Ag^M is singularly uniform and higher than that found anywhere else. Though, as is usual, R_1 ($D^A C^A E^B$) is the commonest Rhesus super-gene, ($D^A C^B E^A$), R_2, is the next most abundant, and nowhere else reaches so high a level: generally 35 to 55 per cent.

The three chief racial groups of America can be distinguished by means of the ABO series. From Mexico southwards, G is greatly predominant. It is even possible that before the advent of the Conquistadores these people were monomorphic for the ABO series. In North American Indians, G is also very common, but, in addition, G^A occurs in varying frequencies, ranging from 1 to over 59 per cent: the latter, found among the Bloods and Blackfoot, being the highest attained in the world. G^B is probably absent, its rare occurrence being almost certainly due to introduction by Europeans. All three allelomorphs are, however, found in Eskimos, and with marked uniformity. G^A occupies about 30 per cent and G^B about 6 per cent of loci.

In addition, it may be added that non-secretors of the G(A) and G(B) antigens are extremely rare among North American Indians; indeed the few secretors among them may be due to racial crossing. Those unable to taste phenyl-thio-urea occupy less than 10 per cent of the population. In some of the tribes this figure reaches 2 per cent of the population (14 per cent of t genes), which is the lowest reported in the world.

Heyerdahl (1952) has developed a detailed theory of the origin of the Polynesians, who include the Maori. This, and the relevant serological evidence, is the subject of a balanced summary by Mourant (1954, pp. 144–7). Heyerdahl holds that Polynesia was populated by two very distinct migrations from America. First, by a fair-haired race, perhaps derived from the Guanches of the Canary Isles who, having reached South America, sailed thence to Polynesia on rafts of balsa logs. Secondly, by migrations of American Indians from the west coast of Canada. Indeed, the still existing Kwatiutl tribe is indicated as their source. This latter contribution appears to have been the more important.

Polynesian serology is little known except from the Maori. They, the Eskimos and the aborigines of North and South America are the only people who have a high frequency of the $(D^A C^B E^A)$ super-gene (R_2) a little under 50 per cent. Indeed, the coastal tribes of British Columbia and the Maori are unique in possessing this as the most frequent Rhesus combination, actually exceeding R_1. In America, Ag^M has a very high frequency, about 80 per cent. The Maori have only about 50 per cent of this gene, but Mourant remarks that 'this is very much more than any non-Polynesian islanders possess'.

In the ABO series, the Maori and the other Polynesians have a very high frequency of G^A, about 40 per cent, but lack G^B. As already mentioned, in South America both these genes are absent. However, in North America G^A is present, and G^B absent. On the Canadian coast, G^A takes a low frequency, 8·5 per cent, but this rises sharply inland, and only a few hundred miles away reaches the highest known value, of 50 to 60 per cent. This close resemblance of the Canadian coastal tribes to the Maori situation is supported by the very high frequency of the P^A gene. However, the reverse situation is shown by the distribution of non-secretors of the G(A) and G(B) antigens, for these are common among the Maori, who must, however, have acquired some Melanesian genes. Owing to the absence of threshold tests, the results of taste-testing the Maori are not decisive. On the whole, therefore, the serological evidence strongly favours a west Canadian origin for the Polynesians, or at least a very considerable contribution from that region.

I have only two additional points to add, arising from my own observations of the Maori in New Zealand, and my discussions there with those who have studied them. First, I was struck with the way in which the Maori fall into two distinct types: respectively narrow-nosed and sharp-featured, broad-nosed and flat-featured. It would be interesting if these could be classified with reasonable clearness, and their blood groups studied separately. Secondly, Dr G. Archey, Director of the Auckland Museum and a distinguished authority on the subject, informed me that probably no pure-bred Maori now remain. All are contaminated by an admixture of European blood. It was only with immense difficulty, about sixty years ago, that a young Maori man and woman, believed completely pure, were obtained in order to serve as models for busts in the Auckland Museum. This fact must evidently be born in mind in discussing the Maori blood groups.

Until recently it was held almost universally that the blood group frequencies were the result of genetic drift and the random survival of certain genotypes in populations reduced to very small numbers at some stage in their existence, or even of racial crossing between those monomorphic for the various genes. The opposite view, that they are maintained by a balance of selective agencies, was favoured almost alone by Sir Ronald Fisher and myself, our arguments being based upon the type of deduction outlined in Chapter 5. Evidence for this concept has in recent years strikingly accumulated (pp. 162-6), so that it cannot now be seriously questioned. It is supported also by the stable blood group ratios of nomadic peoples who, though wandering among other and highly different races, retain for long periods the frequencies characteristic of their origin. This stability is particularly well illustrated by the Gypsies, who carry with them in Europe the distinct serological features of the Hindus of northern India, evidently palaeogenic, p. 180, as shown in the ABO system by their strikingly high group B and low group A frequencies (Table

7). Moreover, their Rhesus super-gene R^2 is exceptionally common while R_2 is rare.

The Jews have of course intermarried in varying degrees with Gentile populations, causing their blood group distributions to differ markedly from one part of the world to another. Their distinct origin is, however, reflected in the fact that in any two regions their serology is more alike than that of the races within which they are immersed. Thus in western Europe they have a higher G^B frequency than Gentiles, one more appropriate to an eastern Mediterranean population (Table 7). This origin is also suggested, even among Ashkenazim (European) Jews, by a high frequency of R_1 (57 per cent) and a low one of the D gene: 33 per cent or slightly less, compared with the usual value of 38 to 45. However, in spite of retaining distinct national characteristics, it is true, as Mourant (1954, p. 70) remarks, that 'the Jews are not a relatively homogeneous population like the more recently dispersed Gypsies'.

Since the blood group frequencies are adaptive, they must be adjusted to given environments, genetic and external. The importance of the genetic environment is shown by such evidence as that derived from the gypsies, by the occurrence of certain isolated races having blood groups very distinct from those of the peoples who surround them (for instance, the Basques), by certain serological similarities maintained over vast and diversified areas (as with the South American aborigines), and by many other features mentioned in the foregoing account. In addition, the adjustment of the blood groups to the external environment is illustrated by the fact that the various, and exceptionally high, values of G^B are characteristic of a region of great-extent in southern Asia.

A few general remarks may be made in summarizing these anthropological facts as a whole. With due caution, it is clearly possible to use the blood groups to trace the impact of one race upon another, whether by gradual merging at their boundaries or by migratory incursion: as with the increasing eastward cline of G^B across Europe into Asia on the one hand, or the tongues of high G^B frequency marking Asiatic invasions into Europe (pp.

168–9) on the other. It is also true that serology provides the best means of identifying isolated groups of primitive peoples (for example, the Berbers), in which it supports other types of anthropological evidence. Yet when we come to consider the facts of world distribution, it is plain that similar serological situations have arisen as independent selective adjustments, and that in consequence it is vain to look for anything approaching the primitive serological conditions of mankind.

Thus it is at first attractive to suggest that an approximate absence of G^B is a primitive characteristic; for the gene is absent both from Australian and South American aborigines (being the two areas of the world where the most ancient types of living organisms are congregated) and from the Basques, who certainly represent a remarkable anthropological survival. However, it is rather common among Negroes and Bushmen, and among Melanesians. The Negroes, furthermore, are one of the two groups who possess G^{A2}, but the Europeans are the other, while an outstanding African quality is the high frequency of the R_O super-gene in which, however, no other primitive people resembles them.

It is at a more recent level, represented for instance by the Polynesian incursion into the Pacific Islands, that the blood groups make their most significant contributions to anthropology. As Darlington (1969, p. 639) remarks, 'there is still time for blood group studies of Eastern Polynesia to answer disputed questions of its (anthropological) history'. But even since those words were written, time is running short owing to the establishment of atomic stations in the region; with the vast incursion of European workers which that involves. Further west and at an earlier date, immense damage was done by the British in bringing to Fiji labourers from India.

However, one point of importance seems frequently missed in discussing serology: it is the fact, to which repeated attention has here been drawn, that genes interact with one another to produce their effects, and form an internal environment to which the blood group frequencies must be adjusted by selection. Blood group genes are often treated as if they take purely additive values when introduced into a population by immigrants, when, in fact, genic

176

interactions may disturb the existing equilibria. Bearing in mind the caution with which they must therefore be treated, the blood groups none the less provide one of the most powerful tools of anthropology.

8.3 PROBLEMS OF PARENTAGE

The application of the laws of Mendel make it possible to state within certain limits which blood groups cannot appear among children of parents whose grouping is known. This type of information can be of critical importance in cases of doubtful paternity, and these are of two principal kinds. Those in which the identity of an infant is in question as, for example, when it is suggested that babies have been accidentally interchanged in a maternity home, and those arising in divorce and kindred proceedings. Thus it may be shown that a child cannot be the offspring of a particular mother or of its legal father; or, alternatively, that a man cannot be the parent of a child attributed to him. Since medical men may be required to give evidence in such cases, they should make sure that they have mastered the genetics of human polymorphism and other relevant conditions.

In all such circumstances, the simple ABO series should first be used on the chance that it may be decisive. For it provides in perfection the requirements for such work: that is to say, there is no difficulty in obtaining anti-sera, the genetic behaviour involved is clearly established, the groups are fully developed at birth, the antigens are stable apparently in all circumstances, and their occurrence is independent of the immediate environment external and internal.

The use of the other blood groups depends upon their varying suitability in the respects just mentioned as so satisfactory for ABO. Some of the necessary sera are as yet rare; while expert opinion is required when systems other than ABO are to be used for paternity tests. The problems then involved are so detailed, and of such a specialized character, that the subject cannot adequately be surveyed here. A brief indication only can be given of the uses and drawbacks of the other blood groups for such work, while further information on the subject may be obtained from Race and Sanger (1968).

The subdivision of group A into groups A1 and A2 can, when carefully handled, provide additional information on paternity. It is generally held, however, that the A1B, A2B distinction, taken alone, is not sufficiently decisive for this purpose. The types of legitimate children excluded from the progeny of any marriage, without and with the subdivision of group A, are listed in Tables 3 and 4 (pp. 149, 151).

The Rhesus groups are quite suitable for paternity tests, for their antigens are well developed at birth. However, some of the anti-sera are at present very rare. Those most likely to be available are anti-C(A), anti-C(B), anti-D(A) and anti-E(A). Race and Sanger, in their second edition (1954), Table 105, list the children possible from all the seventy-eight matings involving the twelve Rhesus groups distinguishable by these four anti-sera.

The MN groups need experienced handling if they are to be employed in paternity tests, for good anti-sera are not easily obtained. The types of children excluded from a given parentage by using them are shown in Table 8. These exclusions are greatly

TABLE 8. Marriages involving the M, N and MN blood groups and the types which cannot appear among their children

Type of marriage	Blood groups absent among children
M × M	N, MN
N × N	M, MN
M × N	M, N
M × MN	N
N × MN	M
MN × MN	—

increased if an anti-L(A) serum be employed in addition. This, however, is not common. Race and Sanger, in their Table 86, give the possible children which can result from each of the twenty-one types of matings involving the MNL groups when anti-L(A), but not anti-L(B) which is at present much too rare to be useful, is available in addition to the two Ag sera.

The P antigen is in general not fully developed at birth and its strength is very variable. Moreover, the Le(A) antigen of the Lewis group does not behave during the first years of life in the way that it does later, for some Le(A—) individuals give a positive reaction against anti-Le(A) sera as infants; furthermore, our knowledge of Le(B) is certainly imperfect. Thus neither the P nor the Lewis series at present appear to be suitable for use in paternity tests.

The remaining blood groups described in Chapter 6 can all be employed for this purpose, in so far as the various anti-sera are available. Since in all of them, absence of a given antibody behaves as a simple recessive, the types of children which they exclude can at once be deduced: that is to say, no positive child can be born to parents who are both negatives, and the occurrence of such indicates illegitimacy.

The reactions of the taste test can be detected in early infancy, so that this also may be used to resolve instances of doubtful paternity. So too can the haptoglobulins (p. 121), as well as the secretion of the ABO substances into the saliva when the individuals belong to groups A, B or AB, but not to O.

The expert witness called upon to explain in Court the evidence on paternity derived from genetics must guard himself against an obvious criticism. He may be asked, 'Is it not true that genetic factors are subject to sudden and unpredictable changes, called mutations, and (granting this fact) is not evidence depending upon their inheritance wholly suspect? How indeed can it be asserted in any legal case that the genetic factor in question has not changed in this way?' The information required to answer these questions has been given in Chapter 3. However, it may be a convenience to provide a correct reply to them here.

The mechanism of heredity in at least all higher organisms is of the particulate or Mendelian type, which depends for its operation upon the extreme permanence of the hereditary factors themselves. Were these to change (mutate) in as many as one individual in even a few thousand, the system would break down. A study of such mutations throughout the widest selection of living organisms demonstrates that a mutation rate of 1 in 50,000 individuals is extremely high, and this frequency is hardly ever exceeded. Inheritance in man is of the same type, and these conclusions are strictly applicable to him. But we are no longer dependent upon analogy for a knowledge of human mutation-rate. It has now been possible to study this in a number of instances (pp. 67–8), in which it proved to be, when most frequent, of the same order of rarity as in other animals (see Neel and Falls, 1951).

It is obvious that, at the frequencies concerned, mutation can be excluded with almost-complete certainty if two unifactorial conditions be studied and both found to be incompatible with the presumed parentage.

8.4 PALAEOGENES AND NEOGENES

Mourant (1954) remarks on the striking diversity in type and frequency of the anti-malaria polymorphisms of man and contrasts it with the marked stability over great areas of most of the blood groups. This distinction has been effectively developed and explained by McWhirter (1967). He points out that organisms will have evolved gene-complexes suited to promote the adaptations for which their ancient major genes have been selected; each one of which therefore operates in a system that will strongly resist minor genetic or evolutionary changes. On the other hand, relatively new major genes, or at least those adjusted to comparatively new situations, will not work in such accurately built-in genetic settings and will prove more sensitive to change. McWhirter has named these two genetic types *palaeogenes* and *neogenes* respectively. Studies on related organisms indicate that palaeogenes with their appropriate gene-complexes are often carried over intact during the evolution of new species and genera. Neogenes, on the other hand, tend to provide less efficient adaptations to relatively recent and less permanent conditions.

We may take examples along the lines suggested by Mourant. There seems good reason to think that Malignant Tertian Malaria, due to *Plasmodium falciparum*, is a disease of fairly recent spread (Darlington, 1964): during perhaps the last 3,500 years. Thus the occurrence of the sickle-celled heterozygotes (pp. 119–20), developed to withstand it, is so easily adjusted that in Africa it quickly responds to public health measures designed to eliminate malaria. Indeed in regions where *P. falciparum* is being success-fully combated, its frequency has dropped from over 20 per cent to 1 or 2 per cent of the population. Against that neogenic situation may be set the palaeogenic one represented by the ABO blood groups of those gypsies who have maintained a closed gene-pool: the striking north Indian characteristics of whom (pp. 169, 174) are still fully retained after centuries of life in Central Europe. In general, we may expect that palaeogenes will often, and neogenes will rarely, be supergenic.

CHAPTER 9

Some Applications of Genetics

9.1 THE USE OF LINKAGE STUDIES

Sex-linked characters have long been known in man; for total sex-linkage is at once evident on account of its characteristic mode of inheritance, while even partial sex-linkage can with no great difficulty be identified owing to the abnormal sex-ratio of affected individuals within particular families. Both these types have already been described in Chapter 2 while an account of autosomal linkage has been given on pp. 15–23.

This is so much the less easy to detect that few instances of it have so far been fully established, apart from the blood group super-genes. There is linkage between the genes for the rare condition of elliptical red blood corpuscles, which are normal in the camel, and the Rhesus super-gene: C.O.V. = 10–15 (Lawler and Sandler, 1954). Also between the gene controlling the nail–patella syndrome and the G locus (of the ABO system), C.O.V. = about 10 (Renwick and Lawler, 1955); between the Lutheran and secretor genes: C.O.V. = 9 (Mohr, 1951); and between the Duffy blood group system and pulverulent cataract, with a C.O.V. of about 2 (Renwick and Lawler, 1963).

Serological polymorphisms and protein variation estimated by means of electrophoresis provide by far the best opportunities for obtaining an adequate map of the human chromosomes in which the marker genes segregate so frequently as to be widely available for linkage tests. Indeed, we are at the dawn of a period in which our own chromosomes can be adequately surveyed. The importance of this, in respect of unifactorial diseases is very evident (see the list of them on pp. 211–18). These are individually rare, but together they constitute an important problem.

Consider, by way of example, a particularly distressing situation

182

such as that presented by Huntington's Chorea. Though appearing occasionally quite early in life, this disease most frequently manifests itself between the ages of 30 and 40: that is to say, not until after the average period for marriage, and even for child-bearing. Since it leads to incapacity and insanity, and since those afflicted transmit it to half their offspring, it is clear that the individuals destined to develop it should not undertake family responsibilities. When a parent suffers from Huntington's Chorea, it is at present impossible to determine to which of his or her children it has been transmitted, except that the chances of any one of them becoming affected, or remaining normal in health and heredity, are equal. As the condition sometimes does not declare itself until the patients are over 60 years of age, it is a particularly sinister aspect of this disease that those offspring of Huntington's Chorea cases who are in reality completely normal at present live under the shadow of becoming insane for the greater part of their lives.

An adequate map of the human chromosomes, based upon polymorphic genes, would largely obviate this disaster, for the linkage relations would make it plain that a given individual was very likely, or very unlikely, to become affected. Since a chromosome map one quarter as complete as that already obtained for one or two animals and plants would suffice for this purpose, it cannot be said that this goal is unreasonably difficult to attain. On the contrary, the extreme degree of polymorphism provided by the blood groups and related phenomena in man makes the necessary information easier to collect in our species than in many lower organisms.

9.2 HEREDITY IN MAN AND ITS DETECTION

The value of correlation in the study of heredity has already been mentioned (pp. 25–6). It was pointed out that the correlation co-efficient indicates the degree to which two variables are associated; such, for instance, as height and eye-colour. If, on the other hand, the association between the same character in different generations be studied, the correlation coefficient then becomes both a method for detecting inheritance and a comparative measure of its intensity. For the standard error of the difference between two correlation coefficients can of course be calculated, so that it can be determined whether the occurrence of a particular condition in one individual is accompanied by its occurrence in a relation more often than in a random sample of the population.

If a character is due to the operation of several genes, its segregation becomes indetectable even when their number is quite small (p. 13). Here the method of correlation largely supersedes that of experimental breeding; but in an organism such as man, in which experiments cannot be undertaken, correlation studies must be applied yet more widely. It is particularly to be noticed that they are applicable to all characters capable of measurement, of whatever kind. They therefore provide a test of the inheritance of mental and moral qualities, in so far as these can be assessed quantitatively. Various other types of numerical comparison are also available for this purpose.

The measurement of intelligence is, of course, a matter of much difficulty, but the carefully devised Binet system seems to provide a just basis for estimating inborn ability in children, totally incapable as I believe it to be of application to intelligent adults. It is so adjusted that environmental and temperamental effects are largely excluded, so that the results obtained by its use are rather nearly proportional to the potential capacity of the children examined. They are expressed as a number, the *Intelligence Quotient* (I.Q.). A valuable example of its use in genetics is provided by Frazer Roberts (1940, pp. 237–8). A group of 3,400 children of school age were given mental tests and classified as Bright (I.Q. > 113), Average (I.Q. = 91–113) and Dull (I.Q. < 91). Three classes were then selected, comprising the brightest 4 per cent, the central 4 per cent and the dullest 8 per cent. The sibs of school age of the children of these three groups were also tested. It was found that 62·3 per cent of the sibs of the brightest children were bright, and only 6·6 per cent were dull; while only 3·7 per cent of the sibs of the dullest children were bright, and 56·3 per cent of them were dull.

Such studies as these have clearly demonstrated that inheritance plays an important part in the development of mental characters. Analysis of a number of them suggest that perhaps 75 per cent of the variation in fundamental intelligence is genetic. It must of course be appreciated that the use any individual subsequently makes of his mental endowments is largely environmental, and quite outside the scope of such tests. It is undoubtedly true also that there is great variation in the rate of development of intellectual, as of physical qualities. Consequently, children of apparently mediocre capacity may occasionally attain to high ability, while many clever children do not fulfil their early promise. Environment is an important factor in such changes. It has also been pointed out that the rate of development of processes in the body is under genetic control (p. 103). We find therefore that some children are, mentally, fast and others slow developers. Some of the latter are, in fact, highly intelligent; indeed they have varied in the direction which has ensured the superior mental qualities of the human race compared with other mammals: prolonging the baby phase and, more relevant here, the period of mental adolescence. Yet at the present time, much less so up to twenty-five years ago, the English educational system is heavily loaded against them. This mistaken arrangement needs to be readjusted in such a fashion that clever but slowly developing children should not be handicapped for the convenience of educational administrators. Furthermore, there is far too much tendency for pressure of work aimed at mere examination passing, rather than real *education*, to dominate school teaching in this country today.

The hereditary component in the control of intelligence is particularly important inasmuch as mating is far from random in respect of mental qualities. There is a strong tendency for the more and the less intelligent members of the community respectively to marry those who approximate to their own mental level.

In general, there can be no doubt that ordinary mental qualities are controlled by normal genetic means, being determined jointly by heredity and environment, as are physical characters. So too are such conditions as tendencies to crime and vagrancy, a fact well demonstrated by the work of Lidbetter (1933) and apparently by the XYY chromosome type (pp. 44–5).

The study of twins plays an important part in the detection of inherited conditions in man. Two types of twins exist, fraternal and identical, and the tendency to produce one or the other of them is inherited. Fraternal twins are due to the simultaneous ovulation of two eggs, both of which are fertilized. They may therefore be either similar or dissimiliar in sex, and they are no more closely related to one another than are ordinary brothers and sisters. Identical twins, on the other hand, arise from a single fertilized egg, which splits into two at an early stage in development. This probably occurs at the first, or very early, cleavage, as suggested by the fact that an instance of differing, but normal, sexes in identical twins was reported at the Tenth International Congress of Genetics. However, mutation apart, these must normally be of the same sex and contain identical genes. The distinction between the two types is usually self-evident. Fraternal twins differ in many respects, while identical twins are almost indistinguishable. Here and there doubtful instances are found, but a thorough examination will always decide to which class they belong: this should include as many physical features, known or presumed to be genetically controlled, as possible. The results can then be confirmed by an examination of the blood groups. These must, of course, all be similar in identical but not in fraternal twins. Consequently, with the number of groupings now available, the distinction between the two types may be detected with fair precision. This subject is, however, beset with complications (see Clarke, 1964, p. 284).

Fraternal twins, though no more alike genetically than ordinary brothers and sisters, tend to live in a more similar environment than they. Statistical studies on the differences between such twins and between normal sibs are therefore of service in determining the effect of environmental variation. Far more important, however, are comparisons between the two classes of twins. The differences between the members of identical pairs are almost entirely environmental; not so those between fraternal ones, for

SOME APPLICATIONS OF GENETICS

these must be as much the outcome of segregation as are the distinctions between normal brothers and sisters. The remarkable similarity, physical and mental, of identical twins demonstrates how large a part of human variation is genetic.

It has been pointed out, however, that owing to the hereditary component in behaviour and in bodily structure, identical twins probably live in a more similar environment than do those of the fraternal type (Macklin, 1940). Thus if one member of an identical pair be physically weak and disinclined to outdoor activities, the other will tend to be so too. Not so in fraternal pairs, whose dissimilar genotypes may give a powerful constitution to one but not to the other; the two being therefore to some extent subject to different surroundings: themselves ultimately the outcome of genetic distinctions. Thus their great variability compared with the identical type may be to some extent accentuated by their environments.

Twin studies are of service in other ways, and these are well illustrated by a useful comparison between the genetics of tuberculosis and pneumonia, drawn by Frazer Roberts (1940). If a number of twins be studied, one of whom suffers from tuberculosis, the frequency with which the other member is also a tuberculosis patient is much higher in the identical than in the fraternal type. Heredity is therefore important in determining susceptibility to the bacillus; but the environment is also a factor, since one member of a pair of identical twins may sometimes be tuberculous when the other is not. Further, in the latter circumstances the prognosis of the disease is particularly hopeful since, as Frazer Roberts points out, the affected individual has succumbed in spite of a high resistance. It is noteworthy also that the site of the lesion is partly determined by hereditary agencies, since this is more often the same in pairs of identical than of fraternal twins. These facts are to be contrasted with the frequency with which pneumonia affects pairs of twins: this is scarcely any higher in the identical than in the fraternal type. Clearly heredity is not an important agent in predisposing to that disease.

When a syphilitic woman gives birth to fraternal twins, it is sometimes found that one member only of them suffers from

187

congenital syphilis. This is strong evidence that susceptibility to it is in part due to the action of genes.

In general, it may be said that, among animals, selection tends to adjust litter size to an optimum; being that which on the average produces the largest number of surviving offspring. Consequently it is to be expected that in man, with a great excess of single births, the mortality of twins should be so high that fewer individuals survive from pairs than from single children. However, Bulmer (1970), who has made a detailed study of this subject, shows that this is not true in England today. But, as he points out, infant mortality has been vastly reduced here in the last hundred and fifty years. Moreover, he shows that it is necessary to take maternal death into account, and this is increased by four and a half times at the birth of twins.

Indeed it seems that during the eighteenth century in Europe, there was little difference in selective value between a twin birth and a single one. Earlier data on the subject are scanty, but Bulmer's investigations indicate that prior to that period the mother of a pair of twins was indeed at a selective disadvantage compared with one who produced a single child.

The occurrence of consanguineous marriages provides valuable material for detecting rare recessive conditions in man. It is a proposition of general application that, with particulate inheritance, inbreeding tends to produce homozygosity. Where self-fertilization occurs, as in plants, the proportion of heterozygotes is actually halved at each generation; for those individuals already homozygous remain so while half the offspring of the heterozygotes are homozygotes. Less close degrees of inbreeding produce homozygosity much less rapidly, but none the less definitely. If an individual is a heterozygote for some rare recessive autosomal condition, the chances that his brother or sister is a heterozygote are $1:1$, they are $1:7$ that a first cousin is one also.

When a recessive affects one individual in 10,000, 2 per cent of the group are heterozygotes (Ford, 1950, pp. 134-6), and if one of these marries an unrelated person the chances against his wife being heterozygous also are $1:49$, but they are $1:7$ if he marries a first cousin. In these circumstances, the chances that the condition shall appear among the offspring are increased sevenfold by a first cousin marriage. If we consider the population as a whole, one individual in 50 is a heterozygote, so that one marriage in 2,500 is capable of producing the homozygous recessive (which appears in one quarter of the offspring), but one first-cousin marriage in 400 can so do: the frequency is 6·25 times increased thereby.

These facts indicate that rare recessives must appear among the offspring of consanguineous marriages far more often than in the general population. The frequent occurrence of consanguinity between the parents of those suffering from a rare disease is a diagnostic sign of much importance. It strongly suggests that the condition is inherited as a simple recessive. In England about 0·6 per cent of all marriages take place between first cousins. Yet over 30 per cent of those suffering from alkaptonuria (p. 104) are the offspring of first-cousin marriages.

Finally, it is necessary to mention a fruitful source of error in the collection of genetic data. If a character is an autosomal 'dominant', half the children who have one affected parent will on the average possess it. But some families will contain too many affected individuals and others will contain too few or none. These in reality cancel each other out but, in recording such family groups, those without affected individuals will be omitted. This will make the condition appear to segregate in excess of the expected equality.

Such an effect is unimportant for genes having heterozygous expression, since the appearance in certain families of a character in approximately half the individuals of successive generations, will make its method of inheritance obvious. But for recessive characters the corresponding error is serious; unless very common, the vast majority of recessives have normal parents, while their expected family incidence is only 1:3. Therefore many families in which both parents are heterozygous will have no affected children, and will be omitted in collecting data. The average frequency per affected family will consequently be much too high, and this is accentuated by the fact that families with many affected members are more likely to be recorded than those with few. Thus it may unjustifiably be doubted if the situation is one of simple unifactorial inheritance. Statistical methods, however, exist which help to reduce these sources of error.

9.3 EUGENICS AND HUMAN RACIAL DISTINCTIONS

As already indicated in the first section of this chapter, the genes responsible for very rare recessive conditions are extremely widespread in the population. Furthermore, it will be recalled that most mutations are disadvantageous in effect, and that unfavourable characters tend to become recessive (Chapter 4). Very large numbers of such genes must therefore exist in man. Consequently we can be sure that each individual will be heterozygous for some of them. It is for this reason only that close inbreeding is generally harmful, so that prohibitions against it exist in the laws of most countries. There is nothing mysterious about the effects of brother and sister marriages, and in an ideally sound stock, one which carries few disadvantageous genes, they would not have undesirable consequences. Their danger is simply due to the generalization already made that inbreeding tends to produce homozygosity, with its consequent segregation of disadvantageous recessives.

These facts indicate how difficult must be any extensive programme of eugenic reform. It is impossible to rid a population of a disease inherited as a simple recessive. Obviously this cannot be done by preventing all who suffer from it from having children, because the heterozygotes are indistinguishable. But even if they could be detected, the gene would be so widely spread that its extermination would be out of the question. Nor is the situation at all simple in respect of the heterozygous conditions (the so-called 'dominants'). These could be eliminated by this means in a single generation, but only in one group of instances is it worth attempting to do this, or rather to lower their incidence by preventing the procreation of at least a proportion of affected persons.

From the point of view of eugenics, heterozygous defects must be grouped into two classes: those which reduce significantly the number of children born to affected individuals, and those which do not. The first of these, of which Darier's Disease is an example, could never be diminished materially in the population by eugenic methods, because the genes controlling them must be distributed with a frequency approximately that of their mutation-rate. Consequently, any one of them would soon be established again at its former level even after its complete elimination. The second group of heterozygous defects, those which have but a slight influence in the number of children born, present a legitimate field for the application of eugenic measures. It includes even a few severe diseases, being those whose first symptoms appear after the average age of reproduction; of these Huntington's Chorea is an instance.

However, L. Darwin (1926) had stressed that the most valuable opportunities for eugenics are provided by multifactorial conditions, which actually allow of more rapid selective improvement. Mental abnormality and deficiency is usually of this type, though it may also be unifactorial. It is of great importance, affecting as it does about 1 per cent of the population; while it tends to increase, rather than to decrease, the number of children born. It is certainly true that a programme of sterilization could materially reduce its incidence.

Of all inherited evils, insanity presents not only one of the most pressing but one of the most difficult problems. For it is easy to damage the working of a delicately balanced system like the human brain in a variety of ways. Thus it comes about that many different genes may all produce apparently similar forms of mental defect. Only here and there, when one of them chances to give rise to some recognizable additional character can the unifactorial nature of a type of mental deficiency be detected, as shown by various instances already mentioned (*e.g.* pp. 67, 104, 183).

It must not be supposed, however, that no advice of a eugenic nature can safely be given. Indeed such advice has already been offered on a number of occasions in this book. Those who belong to a family in which a sex-linked disorder occurs will now know what type of risks they run in marriage. Those who suffer from a defect inherited as a simple 'dominant' must realize that they will transmit the gene responsible for it to half their children: in certain instances they may well question whether they ought to produce offspring. It is particularly important that those who belong to a family in which a dangerous recessive disease has appeared should not marry a near relation. To abstain from so doing is an evident duty. The reason for this will now be clear, and it will be clear also how greatly linkage studies will help those who are faced with such personal problems as these. On the other hand, members of these families can be given a more encouraging piece of information. The conscientious who find that a serious disease is segregating among their relatives and who realize that many of its healthy members must be 'carriers' (that is to say, heterozygotes) may feel not only that they should abstain from marrying a near relative, in which they are quite right, but that they ought to remain celibate in order to avoid disseminating the harmful gene through the population. I have myself encountered this attitude of mind, and know of an individual who rejected the possibility of marriage because of it, while others have allowed it to introduce an additional strain into married life, feeling that by having children they are doing what is morally wrong. Such people can be told that this is a misconception. The genes for rare recessive diseases are so widespread in the community, while they are, in any event, being maintained by mutation which balances their occurrence as homozygotes, that the marriage of heterozygotes is without effect on the general population. It should be realized that if a rare recessive occurs in one out of 100,000 individuals, 0·6 per cent of normal people are carrying the gene for it (see p. 116).

We have so far been occupied with negative 'eugenics', the elimination of disadvantageous characters. There exists also the question of 'positive eugenics', the dissemination of advantageous ones. Among animals such a process is quite practicable, and it has been successfully applied. Close inbreeding, brother and sister mating for example, not only causes the segregation of disadvantageous recessives but, if combined with the elimination of the unfit, purges the race of them. By constant repetition of the process in two lines, rather invariable stocks can be built up each containing relatively few harmful genes. A single 'out-cross' between them ensures much segregation and a wide range of variation, from which the best strains can be selected and maintained by inbreeding.

Such a procedure is obviously inapplicable to man. Moreover, we have at present such poor criteria for deciding what human types are advantageous. Some of us have encountered the conjunction of a weakly body and great brain, and we are all familiar with its depressing converse. Left to human judgment, it is certain that the genius would often never have been born. On a minor scale, a clever man should seek in marriage a partner whose mental powers are good, for we have seen that, on the average, the two will produce intelligent children: but an extensive programme of positive eugenics is one for the distant future.

The question of eugenic reform encroaches upon one of the special preserves of the fanatic, the problem of race and takes our thoughts back to anthropology (pp. 166–77). To some in these days, it has become a creed that such concepts as the British, the Nordic or the Jewish races have, or have not, reality. In this matter, as in many others, extreme views appear to be incorrect.

On the one hand, it is not admissible to maintain that such races are purely artificial. Their characters have some reality, and we have good evidence that they may be preserved when one people is immersed among others. It has already been mentioned that the blood-group frequencies of Jewish populations are more alike than are those of the communities among whom they live (p. 175). On the other hand, particulate inheritance allows the maintenance within each race of vast reserves of variability capable of producing, without racial mixture, far greater diversity than that separating any of them. Recurrent mutation allows the genes of one race to appear in another; and so great a part of their gene-complex have even the most diverse of the human races in common, that no shadow of specific distinction has yet been detected between them (p. 80), though it may well exist. The proportion of genetic material which is different in the more nearly allied races cannot be great. Their special characteristics are, doubtless, preserved by inversions and other chromosomal conditions which hold blocks of genes together.

Ever since the original work of Landsteiner in 1900, it has been known that the blood of one individual injected into another may be either fatal or innocuous. It is therefore a most surprising fact that the blood groups were regarded as of neutral survival value until the contrary was pointed out in 1942 (p. 162). Moreover, it had long been recognized that they provide important criteria in anthropology, but for the wrong reasons.

In discussing this aspect of the blood groups in later years, I was led to remark (Ford, 1957):

> By a curious inversion of logical thought, it was held that their occurrence in distinct and characteristic proportions in the different races of mankind was especially important because the variation involved was selectively neutral. Precisely the contrary is true. The fact that the genes concerned are balanced by selection at optimum frequencies, which differ from race to race, is the one which gives them significance as a criterion of relationship. It does so because in these circumstances their proportions are influenced by the average genotype of the population in which they occur.

This must now be abundantly clear from what has been said so far in regard to serology in general, since we find serological properties polymorphic in one population but not in another; as for example the probable absence of the A and B blood groups in South America before the advent of Europeans, and the approximate absence of blood group B among the Basques. So also with other polymorphic conditions; the gene for red hair, with its additional effects (p. 84) is polymorphic in some regions but absent from others (e.g. in eastern Asia). One may add that a most remarkable feature of racial development is evident in the genetic component of language (pp. 166, 168), demonstrated by the brilliant analysis of Darlington (1947).

The balanced gene-complex necessarily built up in stabilized human populations is responsible for the fact that genes tending to evoke some disease are concentrated in certain races. For example, Fagerhol and Tenfjord (1968) also Fagerhol and Hauge (1968) have shown by an electrophoretic technique that the serum alpha-1-antitrypsin occurs in a considerable number of variants controlled by an allelic series (Pi) of which nine members are known. Some of these differ considerably in their frequencies between different countries. Thus Pi^s is rarer (P < 0·0005) and Pi^f commoner (0·01 < P < 0·025) in Norway than in Spain and Portugal where, however, there is no indication that the Pi^s frequencies differ from one region to another (0·6 < P < 0·7). That allele is also rarer in a number of combined small samples from France, Greece, Italy and Jugoslavia than in Spain (P < 0·0005).

Those homozygous for Pi^z have only about 15 per cent of the normal serum concentration of α_1-antitrypsin, as previously shown by Eriksson (1964). Such people are at a high risk of chronic pulmonary disease, generally leading to death around 50 years of age. Heterozygotes of Pi^z with other alleles do not seem to be affected, so that the condition is recessive. It is indeed probable that there is homozygous disadvantage and heterozygous advantage in this series as a whole. Since the electrophoretic tests are easily performed and can distinguish all the three genotypes, the situation offers evident opportunities for the development of preventative measures.

In a similar way, there are diseases strongly associated with or even confined to particular races. One thinks of the restriction of pentosuria to the Jews or the excess of chorioncarcinoma cases among the Chinese (pp. 99–100). Congenital dislocation of the hip provides a further example of the kind, for it is about three times commoner in Navajo Indians than in the white population of New York City. Clarke (1964) states that two tendencies which dispose towards it are a multifactorial one affecting the construction of the acetabulum and a heterozygous trait producing general laxity of all the joints.

We may of course expect to find that different human races sometimes display minor distinctions controlled by genes adjusted to their particular gene-complexes and giving rise to conditions which would appear highly exceptional or undesirable in another genetic background. Thus Pyke (1963) studied 128 West Indian men and women living in London, half of whom were under 30 years of age and all free from any obvious cardiac symptoms.

Yet in many instances, their electrocardiograms showed profound abnormalities such as would in the English population demonstrate serious heart disease. These effects varied with sex, and in different ways: thus their T waves were inverted in 8 per cent of the men and in 27 per cent of the women, while there was extreme elevation of the ST segment of the tracings in 44 per cent of the men and 6 per cent of the women.

The susceptibility of the races of mankind to infectious diseases depends both upon their gene-complex and upon exposure to the infection. This may promote evolution of a double type: that of the human population and of the causative organism; whether a metazoan parasite, bacterium or virus (Darlington, 1964, p. 138).

For instance, syphilis had become widespread and mild in its clinical effects among the native inhabitants of America, in whom it was endemic. It was brought to the Old World in 1492 by the sailors of Columbus and was carried by the French army into Naples in 1495, whence it swept through Europe with appalling results. The secondary stage was more like an extreme exacerbation of smallpox, so that it was named 'the great pox'. Its decline in virulence, due to relative selection for those more resistant to it, was already noticed within about 50 years, while the change in symptoms which accompanied this adjustment suggested that the Spirochaete itself was also evolving.

But exploration and colonization involve an exchange of diseases, as stressed by Darlington (1969) for, as he points out (pp. 586-7), while 'in fifteen days the Old World received the great pox, in fifteen years the New had acquired measles, tuberculosis, yellow fever and malaria': with, be it noticed, their terrible consequences also in populations which had not been selected for resistance to them. That situation is indeed universally observable. It is well known that Europeans are less susceptible to tuberculosis than are Negroes when living in towns, and in them it runs a more rapid and fulminating course. Almost all the indigenous inhabitants of several of the Pacific islands have been wiped out when a ship carrying measles has called there. At another level, it is likely that the relationship between the blood groups and smallpox (p. 163) is one that has had a profound effect upon the evolution of mankind and the development of human races.

An accurate basis for the race-concept, in so far as this is attainable, should reflect the time available for the accumulation of genetic diversity. It might be reasonable to suggest, for example, that on the average Englishmen had half their ancestors in common thirty generations ago, and that we should have to go back sixty generations before this was true of both Englishmen and Frenchmen. It is certain that the most dissimilar of the races are, judged from this point of view, many times older.

The development of human genetics has advanced greatly in the 30 years since this book was first published. That statement of course applies to our knowledge of the hereditary aspect of disease and of the normal working of the human body; but it has also led to the development of certain concepts of importance for medicine, and indeed for physical anthropology. Thus it is especially to be noticed that the genes interact with one another and with their environment, physiological and external, to produce the effects for which they are responsible: a fact long known, but largely neglected in human genetics even today. Consequently when we assess the impact of a unifactorial, supergenic or multi-factorial condition on one population, we are not to suppose that we can necessarily extrapolate our findings unaltered to another. For instance, data obtained on breast cancer in England and Denmark agree in demonstrating a genetic predisposition to that disease but disagree in associating its familial trend with an earlier onset and with a predisposition to malignancy at other sites (p. 96). But there need be no error here, for the records may reflect a difference in the genetic background of the English and Danish populations. Similarly, when we find a markedly greater liability to smallpox among those of blood groups A and AB, compared with O and B, in India, we are not justified in concluding that this correlation will apply to the same degree elsewhere. It has furthermore been stressed (pp. 176–7) that we cannot necessarily treat the blood group genes as taking purely additive values when introduced into a population by immigrants. Moreover, the analysis presented here clearly excludes Lamarckism, as well as Genetic Drift (save in very small populations), as evolutionary mechanisms.

It is also true that the genetics of the individual have not been sufficiently considered in deciding upon medical prescriptions, for drugs have different quantitative effects in one patient from another. Nor has it been generally recognized that what is pathological in one human race may be normal in another (pp. 195–6). The great genetic variability of individuals, with its interacting effects, leads to another consideration of importance. It indicates the need to test patients for possible allergic reactions to those drugs known on occasion to produce them: the penicillin group and some of those used against malaria, for instance.

It would be wholly misleading to attempt to summarize even this small book, but one aspect of it must surely be stressed in conclusion: that is to say, the importance of polymorphism. It has been shown that this must necessarily be a very common form of variation and that its existence must always indicate the operation of selective forces of importance to the body, however trivial may be the effect by which we recognize its presence in any instance. Here indeed is a situation in which we can make predictions, with a prospect that they will be verified in the future as they have been in the past.

APPENDICES

Cytology

Cytology is the study of cell structure, and I have so far assumed such a knowledge of it as is normally acquired by medical students at an early stage in their career. It may be, however, that some who have no acquaintance with the subject may wish to read this book. Furthermore, those who suspect that their information on mitosis, and especially meiosis, is imperfect, may possibly desire to improve it. I have therefore prepared this brief summary of the relevant aspects of cytology. All who require a more detailed account should consult the fundamental work of Darlington (1937) to whom the logical analysis of cell-division is due; an achievement fundamental to biology.

I. THE RESTING STAGE

The living substance of the body is the *protoplasm*, and this is normally divided into microscopic units, the *cells*. Each cell is composed of two parts, a *nucleus* which controls its activity, and the remaining protoplasm, called *cytoplasm*. The protoplasm of the nucleus, known as *nucleoplasm*, is enclosed in a 'nuclear membrane'.

Tissues grow by a multiplication of their cells, not by an increase in cell size.[1] The nucleus contains the chromosomes, and these carry the hereditary units or *genes* (Chapter 1). Therefore it is a matter of the utmost importance that the chromosomes should be so distributed at cell-division that each of the two new cells into which the old one splits should possess a complete set of them. Normal nuclear division is called *mitosis*. A cell which is not dividing is said to be in the 'resting stage': since it may then be in a state of physiological activity, this term is unfortunate.

The chromosomes are usually, though not quite always, invisible in

[1] In muscle and in the brain, however, increase in bulk does result from an increase in cell size.

the resting nucleus when alive. Cells are generally studied by 'fixing' them (that is to say, killing them with as little distortion as possible) and staining them with dyes which colour their parts differentially. Owing apparently to its high water content, the resting nucleus cannot be examined properly by this means, though we know that the chromosomes always persist in some form from one cell-division to another. Methods of fixing and staining, however, can be made to provide a very accurate picture of dividing nuclei.

As already explained in Chapter I, the chromosomes are present in pairs in every body-cell, amounting to the *diploid number* (2n). One member only of each pair is found in the reproductive cells or *gametes* (the sperm or the egg), which contain the *haploid number* (n).

2. MITOSIS

When a cell divides, it becomes hour-glass shaped, breaks across at the constriction and splits into two. The nucleus, however, has previously divided by mitosis, one of the two products being included in each 'daughter cell'. This is an equating division: that is to say, it ensures that each of the two new cells shall contain the same number of chromosomes, and therefore of genes, as that from which they were derived. We must briefly examine this process. It is somewhat arbitrarily divided into five stages: prophase, prometaphase, metaphase, anaphase and telophase.

Prophase

At the beginning of the prophase, the chromosomes lose water and become denser, so that they can be fixed and stained. It is then found that each has already split *longitudinally* into two parts, the *chromatids*, in preparation for the next cell-division. It will be apparent that the number of chromatids present is the tetraploid (4n). Indirect methods show that the division of the chromosomes takes place towards the end of the resting stage.

It was formerly supposed that a continuous thread, the 'spireme', appeared in the prophase nucleus, and that this subsequently broke up into separate chromosomes. This is incorrect; in reality, the chromosomes always remain quite separate from one another. It is now known that the error was due to imperfect methods of fixation.

Every chromosome possesses a short region, the *spindle attachment* or *centromere*, which does not stain, and here the two chromatids forming each chromosome are still united throughout the prophase. The attachment may occur anywhere except at the extreme end, but its position is constant for any given chromosome.

Prometaphase

In most animals and some plants, a double granule, the *centrosome*, lies outside the nuclear membrane. At the end of the prophase, its two parts separate and move to opposite sides of the nucleus. They remain in connection, however, by a modified region of cytoplasm, the 'spindle'. At the beginning of the prometaphase, the nuclear membrane disappears and the spindle sinks inwards until its axis occupies a line between the two centrosomes. Each chromosome then organizes the nuclear sap into additional spindle elements which coalesce round the original spindle, increasing its bulk. As finally formed therefore, the spindle is composed of two distinct parts.

Structures known as 'spindle fibres', running from the chromosome to each pole, were formerly described. Indeed it was even stated that by their contraction the two chromatids were eventually pulled apart. It is now certain that no such fibres exist. When apparent, they are probably due to the shrinkage of the spindle during fixation, causing cracks to appear between its separately organized nuclear components.

Metaphase

At this stage, the chromosomes lie on the equator of a system whose poles are the centrosomes, and they are 'attached' to the spindle there. That is to say, they are associated with the peripheral spindle elements of nuclear origin which they have themselves organized. It seems that they never come into relation with the central spindle formed of extra-nuclear material which, in fact, does not exist in all forms. The connection of the chromosomes to the spindle is secured only at their spindle attachments, and their long arms stretch out into the cytoplasm. If a chromosome breaks into two parts, only that containing the spindle attachment becomes anchored to the spindle, the other portion floats away and is lost when the cell divides.

The two chromatids forming each chromosome lie even closer together in the metaphase than in the prophase. Consequently, the split separating them is no longer apparent except in transverse section, when the chromosomes assume the shape of figures of eight. Special methods show that the chromatids have the form of a tightly coiled spring within a thin pellicle. At this time, its turns are normally in contact, but they can be separated slightly from one another, so revealing the chromatid structure. The pairs of chromatids are coiled independently of one another. This spiral structure develops during late prophase and the prometaphase. It persists for the remainder of mitosis and, when the resting stage is short, some relic of it may still be visible at the beginning of the next prophase.

Anaphase

The metaphase is a period during which very little change occurs. Usually it is of short duration. The onset of anaphase is marked by the splitting of each spindle attachment into two parts. These then repel one another, so that the two chromatids forming each chromosome are dragged apart in opposite directions towards the poles of the spindle. Thus the prophase chromatids become the anaphase 'daughter chromosomes'. When the two sets have moved some distance from one another, their separation is completed by a change in the equatorial region of the spindle. This elongates, forming a so-called 'stem body', which pushes the two sets of daughter chromosomes yet farther apart.

Telophase

When the two groups of daughter chromosomes have separated widely, they come to rest and undergo a series of changes the reverse of those which took place in the prophase. A nuclear membrane forms, the chromosomes take in water, swell and become increasingly difficult to fix and stain. Finally, they can no longer be demonstrated by this means. The nucleus has then returned to the resting stage.

Either during late anaphase or telophase, the cytoplasm constricts between the two newly formed nuclei, which are thus enclosed within separate cells. Each of them therefore possesses a complete set of undivided chromosomes. These contain the full complement of genes, since they were produced by the longitudinal splitting of chromosomes which had grown to double their original thickness. It will be remembered that this occurred during the previous resting stage. Mitosis is now complete, and the new daughter chromosomes will themselves grow until they also split into chromatids shortly before the next cell-division.

3. MEIOSIS

During the formation of the gametes, an interchange of material takes place between the paternally and maternally derived members of the pairs of homologous chromosomes, resulting in genetic crossing-over, while the chromosome number is reduced from the diploid (2n) to the haploid (n). This is brought about by a special process called meiosis. It consists of two highly modified cell-divisions, being the last which the germ-cells undergo.

Meiosis is defined by Darlington as 'two divisions of the nucleus with but one division of the chromosomes'. Thus it comprises a *first meiotic division* and a *second meiotic division*. These will now briefly be described. Those features, such as spindle formation, which they possess in common with mitosis will be omitted.

(a) THE FIRST MEIOTIC DIVISION

Prophase

The prophase of the first meiotic division is abnormal and complex. It is subdivided into four parts, the leptotene, zygotene, pachytene and diplotene stages.

Leptotene stage. The chromosomes appear as very long, thin threads which are not yet split into chromatids, as they would be at the beginning of a mitotic prophase. A series of darkly staining granules, whose position is constant for a given chromosome, can be detected along their length. These are the *chromomeres*. They are related to the positions of the genes.

Zygotene stage. The homologous chromosomes come together in pairs. This is due to an attraction between pairs of allelomorphs. It does not result from an attraction between whole chromosomes, for these remain unpaired in the region of an inversion (pp. 29–31, 65). As a consequence of this *pairing*, the haploid number (n) of double bodies, or *bivalents*, is formed. These have a split along their length, and they therefore much resemble the chromosomes of a mitotic prophase. The fact that they possess two spindle attachments, however, indicates their very different nature.

Pachytene stage. This comprises the period during which the homologous chromosomes are associated as bivalents. It may last for a considerable time. At first, the members of each bivalent lie parallel, but they soon twist round one another. At the end of pachytene, the chromosomes split into chromatids. As already explained, they normally do so before cell-division starts; but the first meiosis is precocious, for the resting stage preceding it is short or incomplete. Consequently, the chromosomes are not ready to divide until prophase is well advanced.

Diplotene stage. It seems that the force which brings about and maintains chromosome pairing in meiosis is the same as that which keeps together the two chromatids forming each chromosome during the mitotic prophase. This is an attraction of homologous units *in pairs*. But whole chromosomes no longer attract one another after they have split in pachytene, for the attraction is transferred to the pairs of chromatids composing them. Indeed the chromosomes themselves then tend to drift apart, owing to a surface repulsion. However, they are prevented from doing so completely because they are held together at certain points where they have interchanged material. These are the *chiasmata*, and their formation must briefly be described.

FIG. 12. Chiasma formation. A single chiasma has formed between a pair of homologous chromosomes. These are each composed of two chromatids which are still united at the spindle attachment, here represented near one end. It will be seen that the chiasma involves two chromatids, derived from different homologous chromosomes, which interchange material but not partners at a point, while the other two remain intact.

The four chromatids forming the two chromosomes of a homologous pair constitute a *tetrad*. Two of them, derived respectively from the different members of the pair, may interchange material at a point, while the other two remain intact. This interchange of material produces genetic crossing-over. Since it is not accompanied by an interchange of partners, two of the four chromatids must cross over one another in the form of an X, whence the name 'chiasma' (Fig. 12). At least one chiasma nearly always forms between any pair of chromosomes, and the average number may even be as high as eight. However, they do not occur very near together, because the chromatids are stiff enough to resist close twisting. The occurrence of one chiasma therefore prevents the formation of another in its immediate neighbourhood involving the same chromatid. Consequently, double crossing-over does not take place between genes which lie near to one another, whence the phenomenon of 'interference' (p. 21). Complete cytological proof has now been obtained that the chromatids do not merely twist over one another where a chiasma forms, but that they actually interchange material at that point (see Darlington, 1937). Unless this were so, chiasma formation could not provide the physical basis of genetic crossing-over: see also the combined cytological and genetic proof of Stern (1931).

After the chiasmata have formed, the chromatids shorten and thicken as in mitosis, and they rotate through about 180 degrees. Generally also the visible chiasmata (the points at which the chromatids pass over one another to form an X) tend to slide away from the spindle attachment towards the ends of the chromosomes; a change known as *terminalization*. Obviously it does not affect the situation of the interchanged segments, but merely the position of the twist resulting from that interchange.

The prophase is followed by a fairly normal prometaphase.

Metaphase

This resembles the metaphase of a mitosis. It will be realized, however, that each bivalent ('tetrad') possesses two spindle attachments, one paternally and one maternally derived.

Anaphase

The spindle attachments do not now divide, as they do in mitosis. When, therefore, those of homologous chromosomes begin to repel one another, they drag apart pairs of chromatids. Consequent upon chiasma formation, these contain sections both of paternal and of maternal origin.

Both in mitosis and in the first meiosis, the diploid number of bodies move from the equator towards each pole in anaphase. In mitosis, these are the diploid number of undivided chromosomes, but in the first meiosis they are the haploid number of chromosomes divided into pairs of chromatids (which are yet held together at the spindle attachment). The separation of the chromosome sets and their movement towards the poles of the spindle is brought about as in mitosis.

Telophase

This does not differ materially from a normal mitotic telophase. It may or may not be followed by a brief resting stage, the *interphase*.

(b) THE SECOND MEIOSIS

If there has been no interphase, a second meiotic prophase is not needed, and the telophase of the first meiotic division passes directly into the prometaphase of the second. The occurrence of an interphase necessitates a second meiotic prophase, but this is short and exhibits no very noteworthy features.

Prometaphase

This differs from that of a mitosis in two respects only. The chromosomes are of the haploid number, while the chromatids into which each is split are only held together by their spindle attachments; elsewhere they diverge widely.

The metaphase, anaphase and telophase call for no special comment. It will readily be appreciated that they result in the production of gametes containing the haploid number of undivided chromosomes. Occasionally a pair of chromosomes may fail to separate at a meiotic anaphase (non-disjunction). This produces abnormal gametes with, respectively, one chromosome type represented as a pair, when all should be haploid, and another with a chromosome totally absent.

(c) GENERAL SURVEY OF MEIOSIS

In general, it may be said that meiosis consists of two highly abnormal cell-divisions, modified from the mitotic plan. Their evolution has largely been dependent upon an alteration in timing, such that the chromosomes do not split into chromatids until the first prophase is well advanced. This is responsible for chromosome pairing (p. 205).

It should be noticed that, as Darlington originally pointed out, crossing-over with its consequent chiasmata serves two functions. (1) It holds each tetrad together as one set of four chromatids, instead of two sets of two, so that the group can behave in a regular fashion at the first meiotic anaphase. (2) Since it involves an interchange of material between pairs of chromatids derived from different but homologous chromosomes, it brings about crossing-over and so allows recombination between linked genes.

Meiosis also reduces the chromosome number from the diploid to the haploid. It was formerly thought to comprise a single 'reduction division' (the first meiosis) and that this is followed by a mitosis (which corresponds to the second meiosis). Such a view is totally incorrect. The reason for this reveals very clearly the fundamental distinction between mitosis and meiosis, and is therefore worth brief consideration.

The essential function of nuclear division is genetic. A mitosis is an 'equating division', one which ensures that the 'daughter cells' shall each receive the same set of genes as that possessed by the cell from which they were derived. Meiosis is 'reducing' because it brings about Mendelian segregation. If no chiasmata were to arise, the whole of segregation would be effected at the first meiosis, so that the succeeding division (the second meiosis) would be purely equating and therefore mitotic in type. Owing to chiasma formation, however, segregation takes place at the first or the second meiosis alternatively as we pass from one interchanged segment to the next, starting from the spindle attachment. The last cell-division before the gametes are formed is therefore just as much concerned in effecting Mendelian segregation as is the penultimate one. Thus it is a second meiosis, not a mitosis.

Classified List of Some Inherited
Characters in Man

This list is limited to conditions ascertained on fairly good evidence to be determined at least principally by a single main gene. It is not intended to be complete, and in fact it excludes a large number of characters known or presumed to be unifactorial (for information on some of these see the works to which reference is made). In addition, multifactorial inheritance has been studied extensively in man, but the results cannot so usefully be summarized in tabular form.

The characters are arranged in groups, distinguished by letters. Within each of these they are placed in alphabetical order and numbered. This allows cross-references to be given when similar characters are produced by different genes.

Caution must be exercised in using this list. In some instances, environmental agencies may modify the expression of a gene. Further, conditions whose genetics are clearly established may sometimes be inherited in an exceptional way. This may be due to the ordinary gene operating in an unusual gene-complex, or to the action of a distinct gene having superficially similar effects. For example, though albinism is well known to be recessive, dominant albinism has been reported as a great rarity. As already explained (p. 85), no evidence exists that the majority of the so-called 'dominant' characters in man are produced by genes having identical effects in the heterozygous and homozygous states.

Abbreviations are used to give further information on some of these characters and on those which are particularly subject to various irregularities, as follows:

α = Probably some complications.
β = Occasional heterozygous manifestation.
γ = Occasionally not expressed in heterozygote.
δ = Dominance incomplete.
ε = Expression markedly variable.
ζ = Polymorphic.

APPENDIX II

A. BLOOD GROUPS, ζ

The genes controlling the blood groups are not included in this Appendix. They are given at the head of each blood group system in Chapters 6 and 7, and it would involve mere repetition either to list them separately, or to add them to the various sections which follow. Moreover, such a procedure would not be helpful, for the super-genes and multiple allelomorphs involved could not usefully be presented without some genetic details, or apart from the antigens and antibodies which they control, and this would be inappropriate in a mere tabulation of inherited characters and has been done already.

B. SIMPLE RECESSIVES

1. acatalasia (γ).
2. adrenogenital syndrome; great female excess.
3. afibrinogenaemia.
4. albinism.
5. alkaline phosphatase deficiency.
6. alkaptonuria, tends to arthritis.
7. amaurotic idiocy, infantile, common in Jews.
8. amaurotic idiocy, juvenile, rare in Jews.
9. atheroma (C.11).
10. blue or grey eyes (α, ζ).
11. brachymorphia with sperophakia.
12. cystinuria; with another form (β).
13. deaf-mutism.
14. ear-lobes adherent (ζ).
15. epidermolysis bullosa, mild dystrophic (C.34).
16. Fanconi's syndrome (also an environmental form).
17. Fibrocystic disease of pancreas (ζ).
18. freesia, inability to smell (ζ).
19. Friedreich's ataxia.
20. fructosuria.
21. galactosaemia.
22. Gaucher's disease (also dominant form?).
23. von Gierke's disease.
24. Hurler's syndrome (E.8).

25. hyperoxaluria (probably single recessive).
26. hypophosphatasia.
27. ichthyosis congenita.
28. isoniazid, slow inactivation (ζ) subject to modifiers.
29. Laurence–Moon–Biedl syndrome (β, ε).
30. lipoidosis of skin.
31. lipoid proteinosis.
32. McArdle's syndrome (probably recessive).
33. night blindness with myopia (E.13).
34. oligophrenia, recessive (male excess).
35. osteochondrodystrophia (E.15), tends to arthritis.
36. oxalosis (probably recessive).
37. pentosuria (confined to Jews).
38. phenylketonuria (ζ).
39. phenyl-thio-urea, inability to taste (ζ).
40. polydactyly (rarer than 'dominant' form) (C.87).
41. porphyria, congenital.
42. retinitis pigmentosa, with deafness.
43. secretor factor (ζ).
44. spina bifida (C.98).
45. syndactyly (C.103, G.6).

C. SIMPLE 'DOMINANTS' (actually heterozygotes)

1. achondroplastic dwarfing.
2. adenomas, multiple cystic (partly sex-controlled to ♀).
3. albinism, partial often with white forelock.
4. allergy (ζ), the atopic diseases (γ), (♂ excess).
5. anidrotic ectodermal dysphasia (ε), (E.2).
6. aniridia (ε).
7. anomalies of iris (several 'dominants' known), (ε).
8. aplasia cutis congenita.
9. arachnodactyly (ε).
10. arthralgia, periodic.
11. atheroma (γ), (B.9).
12. auditory nerve atrophy.
13. β-aminoisobutyricaciduria (ζ).
14. bone fragility and blue sclerotics.

15. brachydactyly (ε).
16. bullous eruption of feet (γ), (slight ♂ excess).
17. canities praematura.
18. cataract, congenital.
19. cholesterosis cutis.
20. cholinesterase control (ζ), (also D.1).
21. club foot (γ, ε).
22. coloboma, macular.
23. coronary arteriosclerosis (α, ε).
24. Darier's disease.
25. deciduous teeth, fusion (sex-controlled to ♀), (ε).
26. defective enamel of teeth (F.2).
27. diabetes insipidus (γ), (E.6).
28. diabetes, renal.
29. diaphyseal aclasis (partial sex-controlled to ♂), (α).
30. Distrophia myotonica (ε).
31. Ehlers–Danlos syndrome (ε).
32. elliptocytosis (oval erythrocytes).
33. epicanthus.
34. epidermolysis, mild dystrophic (α, γ), (B.15).
35. epidermolysis, simple.
36. epiloia (ε).
37. epithelioma, benign cystic.
38. fingers, apical dystrophy of (ε).
39. fundus dystrophy, 'dominant' form (ε).
40. glaucoma (α, γ), (♂ excess).
41. gout with hypercoricaemia (chiefly sex-controlled to ♂).
42. Habsburg jaw.
43. haemochromatosis, sex-controlled to male.
44. haemophilia–A.
45. haemorrhagic telangiectasia, hereditary.
46. Heberden's nodes (partly sex-controlled to ♀), (α).
47. hip-joint: flattened acetabulum (predisposing to dislocation).
48. Huntington's chorea.
49. ichthyosis vulgaris (γ), (E.9).
50. incisors, absence of (F.6).
51. jaundice, acholuric.
52. keratodermia.

53. labyrinthine deafness (ε).
54. lachrymal duct stenosis.
55. lymphoedema.
56. mandibulo-facial dysostosis (γ, ε).
57. Marfan's syndrome (ε).
58. Marie's cerebellar ataxia.
59. membrane bones, imperfect ossification (ε).
60. microcornea.
61. monilethrix (γ).
62. muscular dystrophy (facio-scapulo-humeral).
63. nails, dystrophy of, with absence of patella (ε).
64. nasal sinus infection, an extreme tendency.
65. nephrosclerosis (manifested also as essential hypertension).
66. neurofibromatoisis, leads to malignancy.
67. night blindness.
68. nystagmus (γ), (F.7).
69. oedema, angioneurotic (non-allergic). ibid. – allergic.
70. opalescent dentine.
71. ophthalmoplegia, progressive.
72. optic nerve atrophy (E.14).
73. Osler's disease.
74. osteoarthritis primary generalized (female excess).
75. osteoarthropathy, tends to arthritis.
76. otosclerosis.
77. ovalocytosis.
78. paralysis, periodic familial (γ).
79. Pelger's anomaly.
80. pelvis, deep acetabulum (partly sex-controlled to ♀).
81. Perthe's disease.
82. Peutz syndrome.
83. phalangeal synostosis.
84. piebalding.
85. pityriasis rubra pilaris.
86. polycystic kidneys, congenital.
87. polydactyly (ε), (B.40).
88. polypi of colon and rectum, multiple (leads to malignancy).
89. porokeratosis (excess of ♂ incidence).
90. porphyria, hepatic, Swedish and South African.

91. premature baldness (ζ, limited to ♂).
92. psoriasis (α, γ).
93. ptosis.
94. retinitis pigmentosa, without deafness (E.17, G.4).
95. retinoblastoma (γ).
96. sebaceous cysts, multiple.
97. spherocytic anaemia (ε).
98. spina bifida (α, γ), (B.44).
99. spastic paraplegia.
100. split hand and foot.
101. squint (convergent and divergent forms, genetically distinct).
102. symphalangism (ε).
103. syndactyly (B.45, G6).
104. teeth, supernumerary.
105. telangiectaria, hereditary.
106. tuberos sclerosis.
107. tylosis palmaris et plantaris (γ), two forms.
108. white forelock (probably dominant in men, recessive in women).
109. v. Willebrand's disease.
110. woolly hair (in Europeans): wavy and curly are perhaps allelo-morphs.
111. xanthomatosis (δ).

D. DISTINGUISHABLE HETEROZYGOTES

1. cholinesterase control (also C.20), (ζ).
2. curly, straight, wavy hair (in Europeans), (α, ζ).
3. haptoglobulin – 1 (ζ).
4. haptoglobulin – 2 (ε, ζ).
5. minor brachydactyly.
6. red hair (ζ).
7. sickling erythrocytes (heterozygotes), sickle-celled anaemia (homo-zygotes) *HbS* (ζ). Another allele, *HbC* (ζ) in West Africa.
8. thalassaemia (ζ), A and B.

E. SEX-LINKED RECESSIVES

1. agammaglobulinaemia.
2. anidrotic ectodermal dysphasia (C.5).
3. choroideraemia (heterozygotes distinguishable by non-pathological symptoms).
4. Christmas disease.
5. colour-blindness, red and green, two loci (β, ζ).
6. diabetes insipidus (C.27).
7. haemophilia (2 allelomorphs).
8. Hurler's syndrome.
9. ichthyosis vulgaris (C.49).
10. 'incomplete' albinism.
11. macula lutea, absence of.
12. muscular dystrophy of childhood.
13. night blindness with myopia (B.33).
14. optic nerve atrophy, hereditary (C.72).
15. osteochondrodystrophia (B.35).
16. potasssium cyanide, inability to smell (ζ).
17. retinitis pigmentosa, without deafness (C.94, G.4).

F. SEX-LINKED 'DOMINANTS'

1. dark-brown eye-colour.
2. defective enamel of teeth, more usually autosomal (C.26).
3. diabetes insipidus, vasopressin resistant.
4. G6PD deficiency (γ, ζ).
5. hypophosphataemia.
6. missing incisors (1 family), (ε), (C.50).
7. nystagmus (only about 30 per cent heterozygous expression) (ε), (C.68).
8. rickets, vitamin-D resistant.

G. PARTIAL SEX-LINKAGE, EVIDENCE DOUBTFUL

1. epidermolysis, severe dystrophic (recessive).
2. Oguchi's disease (recessive).
3. pseudoxanthoma elasticum.

4. retinitis pigmentosa, without deafness (2 allelomorphs, 1 dominant, 1 recessive), (C.94, E.17).
5. spastic paraplegia, recessive.
6. syndactyly (B.45, C.103).
7. total colour-blindness (recessive).
8. xeroderma pigmentosum (recessive).

H. TOTAL SEX-LINKAGE IN Y

1. hairy ears.
2. male sex-determination.

I. HEREDITARY TENDENCIES OF UNCERTAIN NATURE, SOME MULTIFACTORIAL OTHERS UNIFACTORIAL WITH COMPLICATIONS

1. alopecia areata (some clearly 'dominant' and recessive pedigrees known).
2. anaemia, chronic hypochromic (principally ♀).
3. anaemia, pernicious (strong hereditary tendency).
4. cleft palate (distinct from 13).
5. cystinuria.
6. diabetes mellitus (perhaps a multifactorial tendency).
7. essential hypertension (perhaps 'dominant'), (δ).
8. epilepsy, chance against an affected child or sib = 1:20 to 1:40.
9. erythraemia.
10. glaucoma, multifactorial, partly ζ – with one gene of special importance.
11. gout.
12. Grave's disease.
13. hare lip, with or without cleft palate (distinct from 4).
14. hip, congenital dislocation of (ε), female excess.
15. Hirschsprung's disease.
16. hypercholesterolaemia.
17. hypertension, essential (conflicting evidence, whether multifactorial or a single 'dominant').
18. Leber's disease.
19. lung cancer, see pp. 97–8.

20. migraine (possibly 'dominant'), (α).
21. nephritis with deafness, hereditary.
22. paralysis agitans.
23. rheumatoid arthritis, female excess.
24. sclerosis, disseminated.
25. schizophrenia.

References

AIRD, I., BENTALL, H. H. and ROBERTS, J. A. F. (1953) 'A Relationship between Cancer of Stomach and the ABO Blood Groups', *Brit. med. J.* **2**, 799–801.

ALBREY, J. A., *et al.* (1971) 'A New Antibody, Anti-Fy 3, in the Duffy Blood Group System', *Vox Sang.*, **20**, 29–35.

ALLAN, T. M. (1954) 'Fitness, Fertility and the Blood Groups', *Brit. med. J.* **1**, 1437–8, and **2**, 1486.

ALLAN, T. M. (1972) 'ABO Blood Groups and Sex Ratio at Birth', *Brit. med. J.* (27 May), 528.

ALLAN, T. M. and DAWSON, A. A. (1968) 'ABO Blood Groups and Ischaemic Heart Disease in Man', *Brit. Heart J.*, **30**, 377–82.

ALLISON, A. C. (1954) 'Notes on Sickle cell Polymorphism', *Ann. human Genet.*, **19**, 39–57.

ARMALY, M. F. (1967a) 'Inheritance of Dexamethasone Hypertension and Glaucoma', *Archs. Ophthal. N.Y.*, **77**, 747–51.

ARMALY, M. F. (1967b) 'Dexamethasone Ocular Hypertension and Eosinopenia and Glucose Tolerance Test', *Archs. Ophthal. N.Y.*, **78**, 193–7.

AUERBACH, C. and ROBSON, J. M. (1947) 'The Production of Mutations by Chemical Substances', *Proc. Roy. Soc. Edinb.*, B., **62**, 271–83.

BAGSHAWE, K. D., *et al.* (1971) 'ABO Blood Groups in Trophoblastic Neoplasia', *Lancet*, **1**, 553–7.

BARNICOT, N. A. (1950) 'Taste Deficiency for Phenylthiourea in African Negroes and Chinese', *Ann. Eug.*, **15**, 248–54.

BEADLE, G. W. (1960) 'Aspects of Genetics', *Ann. Rev. Physiol.*, **22**, 45–74.

BECKER, B., *et al.* (1966) 'Intraocular Pressure and its Response to Topical Corticosteroids in Diabetics', *Archs. Ophth. N.Y.*, **76**, 477–83.

BELL, J. and HALDANE, J. B. S. (1937) 'The Linkage between

the Genes for Colour-blindness and Haemophilia in Man', *Proc. Roy. Soc. B.* **123**, 119–50.

BENEDICT, F. G. and EMMES, L. E. (1915) 'A Comparison of the Basal Metabolism of Normal Men and Women', *J. Biol. Chem.*, **20**, 253–62.

BENEDICT, F. G. and TALBOT, F. B. (1921) 'Metabolism and Growth from Birth to Puberty', *Carnegie Inst. Wash. Publ.*, **302**, 213.

BERNHARD, W. (1966) 'Über die Beziehung Zwischen ABO-blutgruppen und Pokensterblichkeit in Indien und Pakistan', *Homo* **17**, 111–18.

BOLK, L. (1926) *Das Problem der Menschwerdung*, Jena.

BOYD, W. C. and BOYD, L. G. (1937) 'Sexual and Racial Variations in ability to taste Phenyl-Thio-Carbamide . . .', *Ann. Eug.*, **8**, 46–51.

BULLINI, L. and COLLUZZI, M. (1972) 'Natural Selection and Genetic Drift in protein polymorphism', *Nature*, **239**, 160–1.

BULMER, M. G. (1970) *The Biology of Twinning in Man*, Oxford University Press, Oxford.

(1971), Protein Polymorphism, *Nature*, **234**, 410–11.

CAIN, A. J., KING, J. M. B. and SHEPPARD, P. M. (1960) 'New Data on the Genetics of Polymorphism in the Snail *Cepaea nemoralis*', *Genetics*, **45**, 393–411.

CARTER, C. O., *et al.* (1960) 'Chromosome translocation as a cause of familial mongolism', *Lancet*, **ii**, 678–80.

CHUNG, C. S. and MORTON, N. E. (1961) 'Selection at the ABO locus', *Am. J. hum. Genet.*, **13**, 9–27.

CLARK, W. E. LE GROS (1965) *Fossil evidence for Human Evolution*, University Press, Chicago.

CLARKE, C. A. (1963) 'Further experimental studies on the prevention of Rh haemolytic disease', *Brit. med. J.*, i, 979–84.

(1964, 2nd ed.) *Genetics for the Clinician*, Blackwell Sci. Publ., Oxford.

(1966) 'Prevention of RH Haemolytic Disease', *Vox Sang.* **11**, 641–55.

et al. (1966) 'The dose of gammaglobulin in the prevention of Rh-haemolytic disease', *Brit. med. J.*, i, 213.

(1969) *Selected Topics in Medical Genetics* (pp. 22–44; *ibid.*, with Jones, A. L., pp. 45–68), Oxford Press.

(1970) *Human Genetics and Medicine*, Edward Arnold, London.

(1971*a*) 'Human Genetics' (see p. 181), *The British Encyclopaedia* *of Medical Practice* (chief editor: Sir J. Richardson), Butterworth, London.

(1971*b*) 'Blood Group interactions between Mother and Foetus' in *Ecological Genetics and Evolution* (editor E. R. Creed), Blackwell Sci. Publ., Oxford.

CLARKE, C. A., *et al.* (1955) 'The Relation of the ABO Blood Groups to Duodenal and Gastric Ulceration', *Brit. med. J.*, **2**, 643–6.

CLARKE, C. A., *et al.* (1966) 'Further experimental studies on the prevention of Rh haemolytic disease', *Brit. med. J.*, **1**, 979–84.

CLARKE, C. A., SHEPPARD, P. M. and THORNTON, I. W. B. (1968) 'The Genetics of the Mimetic Butterfly *Papilio memnon*', *Phil. Trans. Roy. Soc. Lond.*, **254**, 37–89.

COCKAYNE, E. A. (1933) *Inherited Abnormalities of the Skin and its Appendages*, Oxford.

CREW, F. A. E. (1937) 'The Sex Ratio', *Rep. Brit. Assn.* (Nottingham), 95–114, London.

CSIK, L. and MATHER, K. (1938) 'The Sex Incidence of certain Hereditary Traits in Man', *Ann. Eug.*, **8**, 126–45.

DARLINGTON, C. D. (1937, 2nd ed.) *Recent Advances in Cytology*, Churchill, London.

(1947) 'The Genetic Component of Language', *Heredity*, **1**, 269–86.

(1956) *Chromosome Botany*, Allen and Unwin, London.

(1958, 2nd ed.) *The Evolution of Genetic Systems*, Oliver and Boyd, Edinburgh.

(1964) *Genetics and Man*, Allen and Unwin, London.

(1968) *The Evolution of Man and Society*, Allen and Unwin, London.

(1969) *The Evolution of Man and Society*, Simon and Schuster, New York.

(1971) 'The Evolution of Polymorphic Systems' in *Ecological Genetics and Evolution* (editor: E. R. Creed), Blackwell Sci. Publ., Oxford.

DARLINGTON, C. D. and MATHER, K. (1949) *The Elements of Genetics*, Allen and Unwin, London.

DARWIN, L. (1926) *The Need for Eugenic Reform*, John Murray, London.

DEMEREC, M. (1948) 'Mutations Induced by Carcinogens', *Brit. J. Cancer*, **2**, 114–17.

DOBZHANSKY, TH. (1951, 3rd ed.) *Genetics and the Origin of Species*, Columbia University Press, New York.

(1966) *Heredity and the Nature of Man*, Signet Books, New York.

ERIKSSON, S. (1964), 'Pulmonary emphysema, and Alpha, Antitrypsin Deficiency', *Acta medica Scandinav.*, **175**, 197–205.

EVANS, D. A. P. (1971) 'Drug Therapy as an Aspect of Ecological Genetics', in *Ecological Genetics and Evolution* (editor: E. R. Creed), Blackwell Sci. Publ., Oxford.

EVANS, D. A. P., MANLEY, K. A. and MCKUSICK, V. A. (1960) 'Genetic control of isoniazid metabolism in Man', *Brit. med. J.*, **2**, 485–91.

FAGERHOL, M. K. and HAUGE, H. E. (1968) 'The Pi Phenotype PM', *Vox Sang.*, **15**, 396–400.

FAGERHOL, M. K. and TENFJORD, O. W. (1968) 'Serum Pi types in some European, American, Asian and African populations', *Acta path. microbiol. Scandinav.*, **72**, 601–8.

FISHER, R. A. (1928) 'The Possible Modification of the Response of the Wild Type to Recurrent Mutation', *Am. Nat.*, **62**, 115–26.

(1930*a*) *The Genetical Theory of Natural Selection*, Oxford.

(1930*b*) 'The Distribution of Gene Ratios for rare Mutations', *Proc. Roy. Soc. Edinb.*, **50**, 204–19.

(1931) 'The Evolution of Dominance', *Biol. Rev.*, **6**, 345–68.

(1935) 'Dominance in Poultry', *Phil. Trans. Roy. Soc. B.*, **225**, 197–226.

(1938) 'Dominance in Poultry', *Proc. Roy. Soc. B.*, **125**, 25–48.

(1958, 13th ed.) *Statistical Methods for Research Workers*, Oliver & Boyd, Edinburgh.

FISHER, R. A., FORD, E. B. and HUXLEY, J. S. (1939) 'Taste-testing the Anthropoid Apes', *Nature*, **144**, 750.

FISHERMAN, E. W. (1960) 'Does the Allergic Diathesis influence Malignancy?', *J. Allergy*, **31**, 74–8.

FORD, E. B. (1931, 1st ed.) *Mendelism and Evolution*, Methuen & Co., London.

(1937) 'Problems of Heredity in the Lepidoptera', *Biol. Rev.*, **12**, 461–503.

(1940*a*) 'Polymorphism and Taxonomy', in *The New Systematics* (editor: J. S. Huxley), Oxford.

(1940*b*) 'Genetic Research in the Lepidoptera', *Ann. Eug.*, **10**, 227–52.

(1942, 1st ed.) *Genetics for Medical Students*, Methuen & Co., London.

(1945) 'Polymorphism', *Biol. Rev.*, **20**, 73–88.

(1949) 'Genetics and Cancer', *Heredity*, **3**, 249–52.

(1950, 2nd ed.) *The Study of Heredity*, Home University Library.

(1955) 'A Uniform Notation for the Human Blood Groups', *Heredity*, **9**, 135–42.

(1957) 'Polymorphism in Plants, Animals and Man', *Nature*, **180**, 1315–9.

(1965) *Genetic Polymorphism*, All Souls Studies, Faber and Faber, London.

(1971, 3rd ed.) *Ecological Genetics*, Chapman and Hall, London.

(1972, 3rd ed.) *Moths*, New Naturalist Series, Collins, London.

FORD, E. B. and HUXLEY, J. S. (1927) 'Mendelian Genes and Rates of Development in *Gammarus chevreuxi*', *Brit. J. Exp. Biol.*, **5**, 112–34.

GAVIN, J. (1964) 'Blood Group Antigen Xg^a in Gibbons', *Nature*, **204**, 1322–3.

GERSHENSON, S. (1928) 'A new Sex-ratio Abnormality in *Drosophila obscura*', *Genetics*, **13**, 488–507.

GLASS, B. (1950) 'The Action of Selection on the Principal *Rh*. Alleles', *Amer. J. hum. Genet.*, **2**, 269–78.

GOLDSCHMIDT, R. (1938) *Physiological Genetics*, New York.

GOODRICH, E. S. (1912) *The Evolution of Living Organisms*, Jack, London.

(1924) *Living Organisms*, Oxford University Press.

GRAHN, D. (1958) 'Acute Radiation Response in Mice from a cross between Radiosensitive and Radioresistant Strains', *Genetics*, **43**, 835–43.

GRUBB, R. and MORGAN, W. T. J. (1949) 'The *Lewis* Blood Group Characters of Erythrocytes and Body Fluids', *Brit. J. exp. Path.*, **30**, 198–208.

GRÜNEBERG, H. (1934) 'The Inheritance of a Disease of the accessory Nasal Cavities', *J. Gen.*, **29**, 367–74.

(1967) 'Sex-linked genes in man and the Lyon hypothesis', *J. Embryol. Exp. Morph.*, **16**, 569–90.

GUNTHER, M. and PENROSE, L. S. (1935) 'The Genetics of Epiloia', *J. Gen.*, **31**, 413–30.

HALDANE, J. B. S.: (1930) 'A Note on Fisher's Theory of the Origin of Dominance', *Am. Nat.*, **64**, 87–90.
(1935) 'The Rate of Spontaneous Mutation of a Human Gene', *J. Gen.*, **31**, 317–26.
(1936) 'A Search for Incomplete Sex-linkage in Man', *Ann. Eug.*, **7**, 28–57.
(1941) 'The Partial Sex-linkage of Recessive Spastic Paraplegia', *J. Gen.*, **41**, 141–7.
(1957) 'The cost of Natural Selection', *J. Gen.*, **55**, 511–24.
HARLAND, S. C. (1937) 'Increased mutability of a gene . . . as a consequence of hybridization', *J. Gen.*, **34**, 153–68.
HARNDEN, D. G., MACLEAN, N. and LANGLANDS, A. O. (1971) 'Carcinoma of the Breast and Klinefelter's syndrome', *J. Medical Genet.*, **8**, 460–1.
HARRIS, H. (1966) 'Enzyme polymorphism in Man', *Proc. Roy. Soc. B*, **164**, 298–310.
(1971) 'Cell fusion and the analysis of malignancy, *Proc. Roy. Soc.*, B, **179**, 1–20.
HARRIS, H., *et al.* (1963) 'Genetic studies on a new variant of serum cholinesterase detected by lectrophoresis', *Ann. Hum. Genet.*, **26**, 359–82.
(1971) 'Polymorphism and Protein Evolution', *J. Med. Genet.*, **8**, 444–52.
HARRIS, R., HARRISON, G. A. and RONDLE, C. J. M. (1963) 'Vaccinia Virus and Human Blood Group A substance', *Acta Genet.* (Basel), **13**, 44–57.
HEYERDAHL, T. (1952) *American Indians in the Pacific*, Allen and Unwin, London.
HOWEL EVANS, W., MC CONNELL, R. B., CLARKE, C. A. and SHEPPARD, P. M. (1958) 'Carcinoma of the oesophagus in association with Keratosis Palmaris et Plantaris (tylosis)', *Q. J. Med.*, **27**, 413–29.
IKIN, E. W. and MOURANT, A. E. (1952) 'The Frequency of the Kidd Blood Group Antigen in Africans', *Man*, **52**, 21.
IKIN, E. W., PRIOR, A. M., RACE, R. R. and TAYLOR, G. L. (1939) 'The Distribution of the A1 A2 BO Blood Groups in England', *Ann. Eug.*, **9**, 409–11.
JACOBS, P. A., *et al.* (1965) 'Aggressive behaviour, mental subnormality and the XYY male', *Nature*, **208**, 1351–2.

REFERENCES

JACOBS, P. A., PRICE, W. H., RICHMOND, S. and RATCLIFF, R. A. W. (1971) 'Chromosome Surveys in Penal Institutions and Approved Schools', *J. Medical Genet.*, **8**, 49–58.

KALLMANN, F. J. (1952) 'Twins and susceptibility to overt male homosexuality', *Amer. J. hum. Genet.*, **4**, 136–46.

KEMP, T. (1948) 'Heredity in Human Cancer', *Brit. J. Cancer*, **2**, 144–9.

KIMURA, M. (1968) 'Evolutionary rate at the molecular level', *Nature*, **217**, 624–6.

KIMURA, M. and OHTA, T. (1971) 'Protein polymorphism as a phase of molecular evolution', *Nature*, **229**, 467–9.

KOLLER, P. C. (1937) The Genetical and Mechanical Properties of the Sex-chromosomes. III Man', *Proc. Roy. Soc. Edinb.*, **57**, 194–214.

KOSSWIG, C. (1929a) 'Uber die veränderte Wirkung von Farbgenen des *Platypoecilus* in der Gattungskreuzung mit *Xiphophorus*', *Z. Indukt, Abstramm.-u. Vererb. Lehre*, **50**, 63–73.

  (1929b) 'Zur Frage der Geschwulstbildung bei Gattungsbastarden der Zahnkarpfen *Xiphophorus* und *Platypoecilus*', *Z. Indukt. Abstamm.-u. Vererb. Lehre*, **52**, 114–20.

LAWLER, S. and SANDLIER, M. (1954) 'Data on linkage in Man', elliptocytosis and blood groups. Families 5, 6 and 7, *Ann. Eugen.*, **18**, 328.

LEHMANN, H. and RYAN, E. (1956) 'The Familial incidence of Low Pseudo-cholinesterase Level', *Lancet*, **2**, 124.

LEHMANN, H., *et al.* (1963) 'Two further Serum Pseudo-cholinesterase Phenotypes as causes of Suxamethonium Apnoea', *Brit. med. J.*, **1**, 1116–18.

LEVENE, R. P. and LEVENE, E. E. (1954) 'The genotypic control of crossing-over in *Drosophila pseudo-obscura*', *Genetics*, **39**, 677–91.

LEWONTIN, R. C. and HUBBY, J. L. (1966) 'The Amount of Variation and Degree of Heterozygosity in natural populations of *Drosophila pseudoobscura*', *Genetics*, **54**, 595–609.

LIDBETTER, E. J. (1933) *Heredity and the Social Problem Group*, London.

LYON, M. F. (1962) 'Sex chromatin and gene action in the mammalian X-chromosome', *Amer. J. hum. Genet.*, **14**, 135–48.

MACKAY, W. D. (1966) 'The incidence of allergic disorders in cancer', *Brit. J. Cancer*, **20**, 434–7.

MACKLIN, M. T. (1940) 'An analysis of Tumours in Monozygous and Dizygous Twins', *J. Hered.* **31**, 277–90.

225

MÄKELÄ, O. (1957) 'Studies in Hemagglutinins in Seeds', *Ann. Med. exp. Fenn.*, **35**, 1–133 (Supplement 11).

MANN, J. D., *et al.* (1962) 'A sex-linked blood-group', *Lancet* (1), 8–10.

MATHER, K. and PHILIP, U. (1940) 'The Inheritance of Hare-lip and Cleft Palate in Man', *Ann. Eug.*, **10**, 403–16.

MATSUNAGA, E. (1962) 'Selective mechanisms operating on ABO and MN blood groups with species reference to prezygotic selection', *Eugen. Quart.*, **9**, 36–43.

MCWHIRTER, K. G. (1967) 'Quantum Genetics of human blood-groups and phoneme-preferences', *Heredity*, **22**, 162–3.

  (1970) 'Ethnic Variations in Incidence of XYY Anomaly', *J. Hered.*, **61**, 192.

MILKMAN, R. D. (1967) 'Heterosis as a major cause of Heterozygosity in Nature', *Genetics*, **55**, 493–5.

MOHR, J. (1951) 'Estimation of linkage between the Lutheran blood group and other hereditary characters', *Acta path. microbiol scand.*, **28**, 207–10.

MOHR, O. L. (1932) 'Woolly hair a dominant mutant Character in Man', *J. Hered.*, **23**, 345–52.

MOURANT, A. E. (1954) *The Distribution of the Human Blood Groups*, Blackwell Science Publications, Oxford.

MULLER, H. J. (1941) 'The Role played by Radiation Mutations in Mankind', *Science*, **93**, 488.

NEEL, J. V. and FALLS, H. F. (1951) 'The rate of mutation of the gene responsible for retinoblastoma in Man', *Science*, **114**, 419–22.

NICHOLLS, E. M. (1969) 'The Genetics of Red Hair', *Human Hered.*, **19**, 36–42.

O'DONALD, P. (1967) 'On the evolution of dominance, over-dominance and balanced polymorphism', *Proc. Roy. Soc. B.*, **168**, 216–28.

PARSONS, P. A. (1958) 'Selection for increased recombination in *Drosophila melanogaster*', *Am. Nat.*, **92**, 255–6.

PARSONS, P. A., MACLEAN, I. T. and LEE, B. T. O. (1969) 'Polymorphism in natural populations controlling radioresistance in *Drosophila*', *Genetics*, **61**, 211–18.

PENROSE, L. S. (1932) 'On the Interaction of Heredity and Environment in the Study of Human Genetics', *J. Gen.*, **25**, 407–22.

  (1935) 'The Detection of Autosomal Linkage . . .', *Ann. Eug.*, **6**, 133–8.

(1950) 'Data for the Study of Linkage in Man', *Ann. Eug.*, **15**, 243–7.

PERRINE, R. P., BROWN, M. J., CLEGG, J. B. and WEATHERALL, D. J. (1972) 'Benign Sickle-cell Anaemia', *Lancet*, **2**, 1163–7.

PIKE, L. A. and DICKENS, A. M. (1954) 'ABO Blood Groups and Toxaemia of Pregnancy', *Brit. med. J.* (2), 321–3.

POPENOE, P. and BROUSSEAU, K. (1932) 'Hereditary Ataxia', *J. Hered.*, **23**, 277–81.

POST, R. H. (1962) 'Population differences in red and green color vision deficiency', *Eugen. Quart.*, **9**, 131–46.

POWELL, J. R. (1971) 'Genetic Polymorphisms in Varied Environments', *Science*, **174**, 1035–6.

PYKE, D. A. (1963) 'Electrocardiographic changes in West Indians', *Proc. Roy. Soc. Med.*, **56**, 567–72.

RACE, R. R. (1965) 'Contributions of blood groups to human genetics', *Proc. Roy. Soc.*, B., **163**, 151–68.

RACE, R. R. and SANGER, R. (1968, 5th ed.) *Blood Groups in Man*, Blackwell Science Publications, Oxford.

RACE, R. R., SANGER, R., LAWLER, S. D., HOLT, E. A. and BERTINSHAW, D. (1949) 'The Lewis Blood Groups of Seventy-nine Families', *Brit. J. exp. Path.*, **30**, 73–83.

RAVEN, R. (1950) 'The Properties and Surgical Problems of Malignant Melanoma', *Ann. roy. Coll. Surgeons*, **6**, 28–55.

REED, T. E., SIMPSON, N. E. and CHOWN, B. (1963) 'The Lyon Hypothesis', *Lancet*, **2**, 467–8.

RENWICK, H. S. and LAWLER, S. D. (1955), 'Genetical linkage between the ABO blood groups and nail-patella loci', *Ann. Eugen.*, **19**, 312.

RENWICK, J. H. and LAWLER, S. D. (1963) 'Probable linkage between a congenital cataract locus and the Duffy blood group locus', *Ann. Hum. Genet.*, **27**, 67–84.

ROBERTS, J. A. FRAZER (1940) *An Introduction to Medical Genetics*, Oxford.

SCHWARTZ, J. T., *et al.* (1972) 'Twin Heritability Study in the Effect of Corticosteroids on Intraocular Pressure', *J. Medical Genet.*, **9**, 137–43.

SHEPPARD, P. M. (1953a) 'Polymorphism and Population Studies', *Symposia Soc. exp. Biol.*, **7**, 274–89.

(1953b) 'Polymorphism, Linkage and the Blood Groups', *Am. Nat.*, 87, 283–94.

SHEPPARD, P. M. and FORD, E. B. (1966) 'Natural selection and the evolution of Dominance', *Heredity*, 21, 139–47.

SINNOTT, E. W., DUNN, L. C. and DOBZHANSKY, T. (1958, 5th ed.) *Principles of Genetics*, McGraw Hill, New York.

SMITH, C. A. B. (1953) 'The Detection of Linkage in Human Genetics', *J. roy. Stat. Soc.*, B., 15, 153–92.

SMITHERS, D. W. (1948) 'Family Histories of 459 Patients with Cancer of the Breast', *Brit. J. Cancer*, 2, 163–7.

STERN, C. (1931) 'Zytologische-genetische Untersuchungen als Beweise für die Morganische Theorie des Faktorenaustanschs', *Biol. Zent.*, 51, 547–87.

STRANDSKOV, H. H. (1948) 'Blood Group Nomenclature', *J. Hered.*, 39, 108–12.

TIPPETT, P. (1967) 'Genetics of the Dombrock Blood Group System', *J. Medical Genet.*, 4, 7–11.

TOKUHATA, G. K. and LILIENFELD, A. H. (1963) 'Familial aggregation of Lung Cancer in Humans', *J. nat. Cancer Inst.*, 30, 289–312.

TURNER, J. R. G. (1969) 'Epistatic selection in the rhesus and MNS (= MNL) blood groups', *Ann. Hum. Genet.*, 33, 197–206.

VOGEL, F. and CHAKRAVARTTI, M. R. (1966) 'ABO Blood Groups and smallpox in a rural Population of West Bengal and Bihar (India)', *Humangenetik*, 3, 166–80.

WEATHERALL, D. J. (1971) 'The Molecular Basis of Thalassaemia', in *Ecological Genetics and Evolution* (editor: E. R. Creed) (pp. 309–23), Blackwell Sci. Publ., Oxford.

WIENER, A. S. (1939, 2nd ed.) *Blood Groups and Blood Transfusion*, London.

(1949) 'Heredity of the *Rh* Blood Types', *Hereditas*, supplementary volume.

WIENER, A. S., ZIEVE, I. and FRIES, J. H. (1936) 'The Inheritance of Allergic Disease', *Ann. Eug.*, 7, 141–78.

WINGE, O. (1932) 'The nature of the sex chromosomes', *Proc. 6th. Int. Congr. Genet.*, 343–55.

WRIGHT, S. (1940) 'The Statistical Consequences of Mendelian Heredity in Relation to Speciation', in *The New Systematics*, Oxford.

Index

Index

INDEX

INDEX

DATE DUE

APR 14 1987			
APR 20 1987			